I0469542

# Why We Have Sex:
## Solving the Darwinian puzzles of sex and speciation

By
**Richard C. Weisenberg**

Text copyright © 2012 Richard C. Weisenberg
**All Rights Reserved**

"Know ye now, Bulkington? Glimpses do ye seem to see of that mortally intolerable truth; that all deep, earnest thinking is but the intrepid effort of the soul to keep the open independence of her sea; while the wildest winds of heaven and earth conspire to cast her on the treacherous, slavish shore? But as in landlessness alone resides highest truth, shoreless, indefinite as God-so better is it to perish in that howling infinite, than be ingloriously dashed upon the lee, even if that were safety! For worm-like, then, oh! who would craven crawl to land!"

Herman Melville


"Doing what little one can to increase the general stock of knowledge is as respectable an object of life, as one can in any likelihood pursue"

Charles Darwin


". . .one of the major challenges for evolutionary theory of the next century to try to put together the different bits and pieces produced by different "schools," particularly concerning levels of selection, chance , necessity and contingency. Only if we are able to play this game, instead of systematically opposing the different possible hypotheses, shall we make significant progress and avoid endless and sterile debates."

Pierre-Henri Gouyon

# Contents

# Introduction

This book originated from my frustration while teaching a course in evolution at Temple University. Evolution is actually not my field of expertise. My degrees are in physics and biophysics, and my research was in cell biology. However, I always had a keen interest in evolutionary biology. This was apparently enough for the chairman of Temple's biology department, who asked me to take over our evolution class when the assigned instructor absconded to another institution. Fortunately, the basics of evolutionary biology are easily mastered, and I think I did a credible job teaching them. The basics were not the problem. The problem was textbook explanations that I found hard to swallow. Two in particular stuck in my craw: The evolution of sex, and the formation of new species.

As a broadly trained scientist, I believe I can detect when a scientific explanation meets reasonable standards of believability. In my estimation, our textbook's explanations of sex and speciation, and in all of the textbooks at that time, failed miserably. The scientific literature on the subject, while more complete, also seemed to fall short. I even went back to Darwin to see what the great man had to say. With regards to sex, he wrote: "We do not even in the least know the final cause of sexuality. . . . The whole subject is as yet hidden in darkness." I think Darwin would hold much the same opinion today.

Darwin, of course, is best known for his classic work, *On the Origin of Species*. Oddly, Darwin had little to say, in the modern sense, about the origin of species. His theory explained how new *varieties* evolved, but not new *species*. Darwin was not sure that species could even be defined. In a letter to a friend, the botanist Joseph Hooker, he wrote:

> It is really laughable to see what different ideas are prominent in various naturalists minds, when they speak of "species"; in some, resemblance is everything and descent of little weight — in some, resemblance

seems to go for nothing, and Creation the reigning idea
— in some, sterility an unfailing test, with others it is
not worth a farthing. It all comes, I believe, from trying
to define the indefinable.

Few modern biologists would accept this view. Darwin, for all his efforts, failed to explain either the origin of species (as defined today) or the evolutionary basis for sex. In essence, he shared my dilemma about sex and speciation. I began to formulate my own views, and attempted to get them published in an appropriate journal, but ran into a stone wall. None of the reviewers claimed that my ideas were wrong, only that they were unproven and lacked quantitative support. This was true, but I was not in a position to provide the missing ingredients. In the end, I decided that it made better sense to introduce these ideas in book form. Perhaps someone with a deeper knowledge of these topics, and with the right skills, will flesh out these ideas.

I also saw the need for a review of the subjects of sex and speciation. Most recent books on these topics were written for specialists, not for students of biology or general, scientifically literate readers. My goal was a book that would be suitable for anyone with an interest in these subjects, and which would also introduce my own ideas. To make this book accessible to the non-specialist I have minimized, as much as possible, the use of technical terms, while maintaining scientific rigor. Nor have I attempted to produce a full review of the relevant scientific literature. This is simply impossible. My apologies in advance to the hundreds of scientists whose articles or books are not cited. (This edition of the book lacks a bibliography, but those who wish to follow up on a topic can email me at rcw@temple.ed. Most citations can be easily found with an internet search.)

This book has four parts. In the first section I describe the present state of knowledge about the formation of new species. While this book is primarily about sex, I believe this topic can only be understood in the context of speciation. In the second section I describe the present state of knowledge about sexual reproduction. Biologists have proposed more than 40 theories to explain the existence of sex, but I will confine my discussion to the main themes. In the third section I offer my own views on these topics.

In brief, I suggest that sex and speciation need to be treated as two parts of a single system. Sex, in my view, is common because it promotes adaptive speciation, while speciation is accelerated by sexual reproduction. Said another way, sex and speciation create a positive feedback loop in which each process promotes the occurrence of the other process. Not only do the birds and the bees do it, the birds and bees, and the millions of other species that occupy the Earth, exist *because* they do it.

For this model to be valid, adaptive speciation must be faster and more robust than it is usually viewed, and I will provide evidence that this is the case. Speciation may have been particularly robust during the early stages in the evolution of sex in single-celled organisms. For these organisms the cost of sex would have been much less than it is in today's multicellular species. As a consequence, sexual organisms rapidly diversified, and sexual reproduction became common among Earth's species. For most of these species, there was no easy path back to an asexual mode of reproduction, even if it would have improved their overall fitness. Because speciation is an important contributor to adaptation, various traits that promote speciation would be selected for, and spread by natural selection. Speciation may proceed faster if reproductive isolation is first imposed on a population by geography, but speciation as an adaptive process is probably common, and sex is a critical element in this process.

In the final section I speculate about the role of sex in human evolution. Here I ask, and attempt to answer, the hypothetical question: Why was Tarzan attracted to the human female, Jane, and not one of the apes he was raised with as a child? The answer to this question goes directly to understanding how humans split off from our primate ancestors and became a new species.

Darwin's *On the Origin of the Species* was published over 150 years ago, and you would think that by now we would have a consensus explanation for how species form and why sex is so common. It is long past time to move forward on these issues. Whatever you think of my ideas, I hope at a minimum you find my discussion of sex and speciation stimulating and informative.

# Chapter 1: The Evolutionary Enigmas of Sex and Speciation

Look out your window. Unless you live in a barren, sandy desert, or an equally barren concrete city, you will see a variety of living organisms. Birds of different species may flit about your yard. Some will be plain in appearance (at least to our eyes), others brightly colored. Many emit beautiful songs. Now ask yourself the following question: How did these different types of birds come to be? By what process did feathered dinosaurs evolve to become finches and robins and the hundreds of other bird species that occupy Earth today?

The short answer is, we don't know. More accurately, scientists cannot agree on how finches and robins became distinct species, and why so many different kinds of birds share the skies. Scientists cannot even agree on how to define the term species, and – amazingly – argue over the reality of species as a meaningful unit of nature.

Here is a more puzzling mystery. All of the birds, and probably all the other animals, that you will see outside of your window reproduce by a process that is inefficient, dangerous and costly. The same process is used by most of the plants in your garden. It is also used by humans. We call this process sexual reproduction.

Some species on this planet reproduce without resorting to sex. These asexual organisms do not spend valuable time and energy seeking a mate, or risk never finding one, or expose themselves to sexually transmitted diseases. In principle, they should reproduce twice as fast as sexual organisms because they only have daughters, each fully able to reproduce without the aid of a male. This is why the reproductive loss inherent to sex is often referred to as the "cost of males." Given the apparent benefits of asexual reproduction, why is sex so common? The short answer is, again, we don't know.

This ignorance is not for a lack of trying. Roughly forty theories have been proposed to explain the prevalence of sex, but none of these are universally accepted. People often look at me in disbelief when I talk about the inability of science to explain either the evolution of species or the existence of sex. Those who know some biology might offer up the standard explanations. The formation of new species, they will say, is based upon geography: "Species form when a group of organisms becomes isolated, such as on an island, and adapts to the new habitat. Later they may mix with the founding species, but will not successfully mate with them." Species, in this view, are a kind of accident of geography.

Sex, on the other hand, must convey significant benefits to overcome its costs. The most common view is that the value of sexual reproduction arises as a consequence of the genetic variation it produces. A typical explanation goes like this: "Sex increases genetic variety in the offspring, which speeds up evolution and gives the offspring a better chance of success."

Both of these answers reflect widely held views among biologists, but both are hotly disputed. In truth, the origin of species and the evolution of sex remain unsolved mysteries. They are "skeletons in the closet" of evolutionary biology. Biologists love to tell you about the successes of evolutionary theory, which are many, but they cannot agree how evolution led to two of the most dramatic aspects of life on Earth: Its diversity, and the fact that so many organisms reproduce sexually. Darwin was also puzzled by these two issues, but never found a satisfactory explanation for either of them.

## Darwin's first dilemma

Given the title of Darwin's greatest work, one would think that he devoted a good deal of effort to the origin of species. One would be wrong in this assessment. With respect to speciation, Darwin faced a dilemma. He saw that natural selection would lead to the spread of new varieties, but this would occur slowly, through gradual change. Early forms would segue into more advanced forms, leaving transitional stages along the way. Nothing in his theory seemed to predict the evolution of discrete types, or species. He expressed the problem this way:

> As on the theory of natural selection an interminable number of intermediate forms must have existed, linking together all the species in each group by gradations as fine as our present varieties, it may be asked, why do we not see these linking forms all around us? Why are not all organic beings blended together in an inextricable chaos?

Darwin attempted to explain the missing "inextricable chaos":

> For we have reason to believe that only a few species are undergoing change at any one period; and all changes are slowly effected. I have also shown that the intermediate varieties which will at first probably exist in the intermediate zones, will be liable to be supplanted by the allied forms on either hand; and the latter, from existing in greater numbers, will generally be modified and improved at a quicker rate than the intermediate varieties, which exist in lesser numbers; so that the intermediate varieties will, in the long run, be supplanted and exterminated.

The transitional forms, in Darwin's view, had merely disappeared. Darwin's attempt to explain why these transitional forms became extinct must have seemed lacking in substance even to him, and this explanation has not survived the test of time. When modern biologists discuss speciation they rarely cite Darwin's explanation, perhaps as a courtesy.

Darwin also differed from modern biologists in his view of the nature of species. He considered species to differ only in degree from varieties:

> I look at the term species as one arbitrarily given for the sake of convenience to a set of individuals closely resembling each other . . . it does not essentially differ from the term variety, which is given to less distinct and more fluctuating forms. The term variety, again in comparison with mere individual difference, is also applied arbitrarily, and for mere convenience sake.

Thus, a finch and a robin, to Darwin, merely reflect extremes of the differences present within a flock of either type. Darwin must have known, however, that this idea was untenable. Hybrid sterility, for example, is hard to explain if species differ from varieties only in degree. It is clear from his writings that Darwin shared the commonsense view that species are distinct natural entities. In the end, Darwin abandoned his attempts to explain how new species emerged, calling it the "mystery of mysteries" of evolution.

Today, most biologists divide into two camps: Those who believe geographical isolation (called "allopatry" is essential for speciation and those who believe speciation can occur without isolation ("sympatry"). About the only thing these groups agree on is that the other side is badly mistaken. To be fair, some biologists have assumed a more neutral stance. For example, in 2009 the journal Nature published a summary of ideas on speciation in the form of a series of questions and answers (the author of this piece, Andrew Hendry of the McGill University, was answering his own questions). To the question "Does speciation require geographical isolation (allopatry)?" he had this to say: "Yes. Wait. I mean no. Or perhaps maybe . . ."

This may be one of the more coherent answers that I have seen in response to this question. Of course, the point of Hendry's answer was to emphasize the ongoing uncertainty about the role of geographical isolation in speciation. Some biologists consider a belief in sympatric speciation akin to scientific heresy. The eminent evolutionary biologist Ernst Mayr expressed it this way in 1963: "One would think that it should no longer be necessary to devote much time to this topic, but past experience permits one to predict that the issue will be raised again at regular intervals. Sympatric speciation is like the Lernaean Hydra which grew two new heads whenever one of its old heads was cut off." Another eminent evolutionary biologist of his time, Theodosius Dobzhansky, concurred: "Sympatric speciation is like the measles; everyone gets it and we all get over it." Both Mayr and Dobzhansky had a huge impact on evolutionary thought for much of the twentieth century, and their views became almost universally accepted.

More recently, some theorists have argued that sympatric speciation *might* happen under limited and rare circumstances. A few (including this writer) take the stance that sympatric speciation is common, and may be directly linked to Darwin's second dilemma, sexual reproduction.

## Darwin's second dilemma

The other dilemma that Darwin faced concerned sex. Much of the *Origin of the Species* is devoted to sex in one form or another. Sexual selection and hybrid sterility get their own chapters, and his later works are even more focused on the biology of sex. In spite of this fascination with all things sexual, Darwin never attempts in his published works to explain why sex exists in the first place. Only in his notebooks did Darwin propose a solution:

> Without sexual crossing, there would be endless changes, & hence no feature would be deeply impressed on it, & hence there could not be improvement, & hence not in higher animals — it was absolutely necessary that Physical changes should act not on individuals, but on masses of individuals. — so that the changes should be slow & bear relation to the whole changes of country, & not to the local changes — this could only be effected by sexes.

The lack of clarity in this explanation is understandable, as it is only a note by Darwin to himself. That he never expanded on this view suggests that he found the argument to be weak, and it never reappears in his published works. Instead of providing a plausible theory, he eventually, in 1868, made this confession: "We do not even in the least know the final cause of sexuality. . . . The whole subject is as yet hidden in darkness."

Darwin's solution to the dilemma of sex was to admit defeat.

Today, biologists still struggle to understand the reasons for sex, and have yet to reach a consensus. Indeed, the level of disagreement may be unmatched in science. You might think that biologists agree on the basic issues, but in this you would be mistaken. As one well-respected researcher expressed it: "We have the answers. We cannot agree on them, that is all." Most explanations of sex focus on its ability to produce genetic variation. Yet in 2007 Henry Heng of Wayne State University wrote the following: "According to traditional perceptions, asexual reproduction should lead to more precise offspring and sexual reproduction to more diverse offspring. In reality, however, the relationship is quite the opposite." Thus, if Heng is right, nearly all theories on the evolution of sex have got things backwards!

Sex remains "hidden in darkness," but scientists have plenty of theories that they hope will cast light on the topic. Each of many explanations that have been proposed makes a bit of sense, at least to their authors, and any one of them might be valid under certain ecological conditions. No single theory, however, seems to explain the prevalence of sex among Earth's species. As an alternative, some researchers have proposed that sex might provide multiple benefits, and that these act in combination to help maintain sexual reproduction. Most scientists, I believe, would expect a phenomenon as widespread and important as sex to have a single cause. Having multiple theories of sex is a bit like having multiple explanations of why bricks, feathers and apples all fall downward. The search for this single explanation continues.

The existence of so many theories has become an issue itself, as demonstrated by a 2010 article in the Journal of Heredity. The title of the article, by Stephanie Meirmans and Roger Strand of the University of Bergen, speaks for itself: "Why are there so many theories for sex, and what do we do with them?"

What indeed?

## Theories of sex, a brief overview

As part of the "research" for this book I asked people, both scientists and nonscientists, why sex exists. The most common answer was "for reproduction." Next was "to create genetic variety." A few answered "for pleasure." The first and last of these responses can be quickly discarded. Sex is often a precursor to reproduction, but it is not a means of reproduction *per se*, as asexual organisms reproduce just fine. The pleasure we associate with sex is, of course, to encourage us to have sex, not the reason for it.

Among evolutionary scientists, the creation of genetic diversity provides the most widely accepted explanation for sex. These scientists cannot, however, agree on why genetic diversity is of value. One view is that sex allows organisms to adapt more rapidly to a changing ecosystem, or be more efficient at utilizing a complex one. When an ecosystem changes, asexual organisms must wait for new mutations that allow them to adapt to the new conditions. Sexual organisms, by constantly creating new genetic combinations are more responsive to a changeable or complex environment.

This class of theories may seem limited in scope because they assume the existence of rapidly changing environmental conditions. However, even a stable ecosystem is likely to undergo periodic change. A popular theory based on this fact is called the Red Queen, named after the character from Alice in Wonderland who is always running, but never gets anywhere. As she explains it: "It takes all the running you can do, to keep in the same place." The Red Queen model views nature as a kind of arms race between predator and prey, or parasites and hosts. Predators constantly evolve to improve their chances at catching prey, while the prey organisms evolve in turn to improve their chances of not being eaten. Sex, by increasing genetic diversity, gives offspring a chance to stay a step ahead of its enemies (or its food source).

Another popular theory concerns the impact of harmful mutations, and how sex can mitigate these. Theories of this type view sex as a sort of genetic eraser, which can limit the impact of harmful mutations. Several other ideas have been proposed, but most of these get little attention. None of the proposed models have gained anything like universal acceptance. Even the most widely accepted model, the Red Queen, has come under theoretical and experimental challenge, but no other ideas appear to have taken over its dominant position. I will discuss these models in more detail in later chapters, but first I want to suggest the outlines of a possible solution.

# CHAPTER 2: Looking for a solution

When it comes to the science of sex and speciation, confusion reigns. Theories and definitions abound, and little hard data exists to choose among them. To make this book readable I will focus on a few core ideas, although I will attempt to summarize as many others as possible. To this end, I will use a single definition of species, and focus on only a few of the many theories of sex that have been proposed. Even so, as I get into the details, eyes may glaze over, and pages may get scanned, or skipped all together. Before that happens I want to lay out the basic ideas behind this book.

While my main focus in this book is sex, I believe that we cannot understand this topic without a firm understanding of speciation. My reasons for this assertion will become clear in due course.

## The puzzle of speciation

The science of sex may be split between dozens of theories, but the science of speciation is essentially a battle over the role of geographic isolation. For the sake of simplicity, I limit discussion to the two extremes; compete isolation, called allopatry, versus no isolation, called sympatry. These terms mean, respectively, "different fatherland," and "same fatherland," but should not be viewed as rigid, as isolation does not always fit either of these extremes, and other modes of speciation (called peripatric, parapatric and hetropatric) are also possible. The battle that rages, however, is about sympatric speciation, and that will be my focus.

The lever of disrespect that has been shown to sympatric speciation is astounding, but in recent years has abated. This does not mean that sympatric speciation has been allowed out of the biologist's dog house. A major 2004 book on the topic (*Speciation*, by Jerry Coyn and H. Allen Orr) takes a firm stand on the issue. The authors argue that allopatric speciation is the default position, or null hypothesis, and allopatric speciation is to be accepted unless it can be clearly excluded. Coyne and Orr note that "Some may find this strategy unacceptable ..."

Among those who found this approach unacceptable was Hope Hollocher, an evolutionary biologist at Notre Dame University, who wrote, ". . . in allopatry, species remain undefined by the [Biological Species Concept] because they are not in contact. Yet, Coyne and Orr claim that most speciation occurs in allopatry. Logically, you cannot have it both ways ... You cannot have the most common geographical mode of speciation lead to species groups that remain undefined with that context."

More recently, in the September 2012 issue of Scientific American, which focused on where science is headed in next 100,000 years, Jerry Coyne wrote the following: "Most new species in nature appear when a population becomes geographically isolated from other populations." This seems to allow little room for debate. Yet later in the same article he makes the following observation: "For geographically separated species of *Drosophila*, we found that the two types of barriers – mating problems and sterile offspring – evolve at about the same rate. But for species *cohabiting the same area* [my emphasis], interbreeding barriers seem to evolve quicker." Apparently geographic isolation is required for speciation – except when it's not.

While evolutionary biologists continue to debate the issue, textbooks typically reflect the consensus view that allopatric speciation dominates. For example, the first textbook I used to teach evolution had as its example of speciation the cichlids of the Rift Valley Lakes of Africa. These large, ancient lakes contain hundreds of species of these colorful fish. At one time, according to the text, a single large lake was present, and contained just a single species of cichlid. This lake supposedly dried up, forming several smaller lakes and ponds. Within these isolated bodies of water, the cichlids began to evolve into distinct types. Later, the rains came again, and the lake refilled. By this time the cichlids from the smaller lakes had become sufficiently different that they would no longer mate with the original type, or if they did mate the hybrids were poorly adapted and did not survive.

The result, according to this model, was the formation of new, distinct species. As a teacher I felt obliged to present the concept of allopatric speciation as clearly as I could, but this was a problem as I found the textbook's explanation to be weak. Actually, I considered it to be little more than scientific hand-waving. Even if the proposed cycles of emptying and filling of the lake could be proven, the smaller lakes would remain connected by streams that would allow mixing of the fish populations, and periods of heavy rains would have caused floods and occasional merging of the smaller lakes.

Even more troubling was the assumption that geographical isolation was, by itself, sufficient to lead to new species formation. This explanation required that the ecology of the smaller lakes differed enough to cause the evolution of new adaptations. That seemed highly unlikely. Nor, was it clear why the different populations should avoid mating with each other when they came together again. (Those of you with a background in evolution might be shouting "secondary reinforcement" as you read this, and I promise to deal with that later.)

The next textbook I used focused on the speciation of the finches on the Galapagos Islands. This group of birds, referred to collectively as Darwin's Finches, consists of several distinct species on each of the islands. The large island of Isabella, for example, has 13 named species of finches. The existence of several species on an island was explained as follows: Sometime after the Galapagos Islands formed a few finches blew over from the mainland of South America. They thrived in the new environment, which was rich in food and lacking predators. These birds subsequently colonized several other islands, where they evolved new adaptations and different mating habits. Eventually, some birds returned to the originating island, and established themselves as new species.

For this model to be valid, after the birds dispersed to distant islands they had to remain isolated long enough to evolve distinct adaptations – for no particular reason – and then, and only then, make their way back to the original island. Once there, they must find their ancestral relatives to be sexually unattractive, and proceed to evolve into a distinct species. Again, no direct evidence existed in support of this model. Nor did it make much sense. It required, among other assumptions, that the different islands were ecologically distinct. They are, but they all contain seed-producing plants and a variety of insects. The model never explains why the birds of one island would have concentrated on eating insects, while they focused on seeds on another island.

What's going on here? Why did two popular evolution textbooks use such weak examples to illustrate speciation? The answer, I suppose, is that they had nothing better to offer. For decades the prevailing explanation of speciation was that it required geographic isolation. This is why lakes and islands so often come into play when biologists discuss speciation.

To me, however, allopatric speciation never made sense, as it seemed to have a fatal weakness: If isolation is a precondition to speciation, then the less mobile a species is, the more likely it would split into new species. The opposite seemed to be true. A population of mice, for example, can be isolated by a river or a highway. Yet they do not become new species. Birds, by comparison, are hard to keep in one place. They can fly, after all. Yet many more species of birds exist than species of mammals. Something else had to be going on.

Allopatric speciation appears to have become the default position for one primary reason – existing mathematical theories "proved" that sympatric speciation was slow, or even impossible. These theories, in my view, are akin to those that "proved" that bumblebees cannot fly. Bumblebees fly, and sympatric speciation, in my estimation, is both possible and common. Even the most avid supporter of allopatric models should accept this as fact. Why? Because this theory must also explain what happens when isolated populations come back into contact. This phenomenon, called secondary reinforcement, is basically sympatric speciation that occurs following a spell of allopatry. I will have much more to say about sympatric speciation, but for the moment I leave you with the following thought: Do we actually need multiple theories of speciation?   An oddity of allopatric theories is that they only become interesting when the new and ancestral species come back into contact – that is when they become sympatric!

The literature on speciation is as dense and impenetrable as an Amazonian jungle. To me one fact stands out: Speciation is closely tied to sex. The arguments over speciation are about how this process occurs in *sexual* organisms. Speciation in asexuals is an entirely separate topic, and some have argued that the term species should not even be applied to asexuals.

While I taught my students about allopatric speciation, I felt obligated to inform them that I did not believe that it could account for most species. I believed that sympatric speciation must be possible, but struggled to explain to my students how speciation could occur without an imposed period of reproductive isolation. Fortunately, for me and my students, a pair of articles appeared in 1999 that provided realistic theories for the alternative of sympatric speciation. These articles considered how speciation might occur within a single, interbreeding population, and came up with essentially the same solution.

According to the widely held view at that time, sympatric speciation should be inhibited by gene flow within an interbreeding population. Ongoing hybridization between a new emerging species with the preexisting species would prevent their genetic separation.  For sympatric speciation to occur a mechanism was needed that would limit gene flow between the old and new species. In allopatric speciation this is accomplished by geography, but it was not clear how this could be accomplished in a sympatric population.

The 1999 articles found a solution to the hybridization issue. I will focus on an article by Alexey Kondrashov and Fyodor Kondrashov. The other article, by Ulf Dieckmann and Michael Doebeli is different in detail, but is based upon the same insight. Basically, the Kondrashovs realized that previous models of speciation had focused on traits controlled by single genes. This made the math simple, but was biologically unrealistic. Many important traits are under the influence of multiple genes. For example, body size in people is influenced by about 100 different genes. Another well-known example of this phenomenon is human skin pigmentation, which is controlled by about seven genes. One consequence of this fact is that people come in many different sizes and skin tones. Body height and skin pigmentation are examples of "quantitative traits," which are distinguished from a qualitative trait, such as pea flower color, which comes in distinct forms.

A quantitative trait is determined, approximately, by the combined impact of the several genes that control that trait. As a consequence, offspring tend to display the average value of their parent's quantitative traits. Quantitative traits provide a kind of "truth in advertising" in that what you see, genetically speaking, is what you get. This distinguishes them from qualitative traits, such as flower color in pea plants, where a dominant gene may hide the presence of a recessive one.

The Kondrashovs considered a hypothetical fish in which a trait, body size, was controlled by several genes. The fish were allowed to evolve in an environment in which two kinds of prey existed, one large and one small. This condition favored fish that were either large or small, but selected against fish that were intermediate in size. Females were assumed to choose a mate according to his color, which varied from red to blue within the population. Some females preferred red in a mate, and some blue.

By chance, a large red-preferring female may choose a red mate who is also large. Their offspring will tend to be large, and have genes for red preference and color. Because of their size, these offspring will be successful, and have offspring of their own. After several generations, fish evolve that are distinguished by their large size and by their mating habits. In this way a new species of fish evolves, without a period of geographic isolation.

Biologists continue to debate the significance of sympatric speciation – and that is ok. What is important for the present discussion is that *some* species are the result of sympatric speciation, and that for *some* of these species the mechanism of speciation is basically that described by the models of Kondrashov and Kondrashov and Dieckmann and Doebeli.

Scientists can, and certainly will, argue about how common this kind of speciation is, but I suspect that it was common enough to have had an important consequence – it may have made sex valuable enough to dominate the Earth's species.

## The function of species

After a few years of teaching evolution I began to feel comfortable with the topic, and at least shook off the fear that I was depriving my students of a proper education. When I had to explain sex and speciation, however, no amount of preparation could eliminate the feeling that my students were being cheated. As I struggled to resolve these issues, I began to develop my own views. I came to believe that the fundamental problem was treating sex and speciation as separate topics. Perhaps they were really two aspects of a single topic.

It's not that these topics are never treated together. Discussions of speciation, for example, invariably include the role of sexual reproduction. To even define the term species requires a consideration of reproductive behavior. By comparison, the literature on sex is nearly devoid of any consideration of speciation. Only two investigators, Leonard Nunny in 1989 and S. M. Stanley in 1975, appear to have given much thought to the linkage between sex and speciation, and their articles on the subject are rarely cited. I will have more to say about their ideas later.

One fact, in particular, struck me as significant. Speciation, whether allopatric or sympatric, is usually viewed as a *consequence* of adaptation. According to this view, a population adapts to an ecological state by gradual change until it becomes sufficiently different to be designated as a new species. However, speciation may more properly be seen as a *means* of adaption. For example, in the Kondrashovs' example a fish species adapts to the presence of large and small prey through speciation. However, what is important, and what has been little noted, is that if speciation is a means of adaptation, then any process that promotes speciation is itself a means of adaptation, and will be selected for.

Sex is such a process.

Evolutionary scientists have had more than 150 years to find a solution to the problem of sex. Without success. It is either a very hard problem – or they have been barking up the wrong theoretical tree. Consider the basic issue that has guided research on sex. Biologists have focused on finding reasons why sex is so common in the face of its costs. Sara Otto, of the University of Toronto, stated the issue this way: "In the face of such legendary costs, we might expect sexual reproduction to be rare. Yet, the vast majority of eukaryotic organisms reproduce sexually—at least occasionally. Among named animal species, only ~0.1% are considered to be exclusively asexual . . ."

Simple logic appears to tell us that sex must be valuable because 99.9% of animal species (and about 99% of plants) reproduce sexually. However, this is not the only possible logical relationship that can be drawn from this fact. One could also conclude from this correlation that *sex causes speciation*. By this logic, species are mostly sexual for the simple reason that they are a product of sexual reproduction. Perhaps what is important is not that the birds and the bees do it, but that the birds and the bees, and the millions of other species that occupy the Earth, exist *because* they do it.

Most of the time, cause and effect relationships in nature seem clear. We know, for example, that wet pavements do not cause rain. Likewise, if deformed frogs are found in a pond, and the pond contains a toxic chemical, we would deduce that the toxin caused the deformation of the frogs. Logically, however, such a correlation is also consistent with a hypothesis that deformed frogs caused the pond to be contaminated with the toxin! This seems unlikely, of course, and few scientists would treat it as a viable hypothesis.

Causation is not always obvious. My favorite example of confused causality comes from observations that students who choose to sit in the front of the classroom tend to get better grades than those in the back. This could mean either that closeness to the teacher leads to improved learning, or that better students tend to sit in the front of the classroom. Only carefully designed experiments could resolve which of these explanations is most valid.

Popular theories of sex begin with the assumption that sex must be beneficial because most species use it, but the logic here may be backwards. Perhaps most species are sexual because sex leads to speciation. This hypothesis has received very little attention, which is understandable. It requires, if valid, that we drastically alter the standard views of both sex and speciation. One would need a good reason to do this.

Consider the deformed frogs again. Suppose scientists test the hypothesis that a toxin found in a pond causes frogs to become deformed. Years of experimentation, however, produce only negative results. Finally, out of desperation, they look at the alternative causal relationship. Much to their surprise, and chagrin, they discover that frogs are subject to a deformity-causing bacterial infection. Frogs respond to an infection by producing an antibacterial toxin. The toxin in the pond is not the cause of the frog's deformities, but a result of them!

For years, biologists have struggled to explain why sex is beneficial to species that use it . . .

Oops! I appear to have just made a classic mistake in evolutionary biology – the kind of thing I would jump on my students for doing. As a rule, evolution works on individuals, and is driven by the reproductive benefits that they gain from having certain genes. Evolution does not usually promote the "survival of the species" (although I will discuss an important exception later). This kind of statement you might find in a poorly researched nature show on TV, not in a serious book on evolution. Two rams bashing their heads together are not helping to improve their species. They are fighting for a chance to reproduce. I need to be more careful in my choice of words. Let me rephrase:

For years, biologists have struggled to explain why sex is beneficial to organisms that constitute a species.

Organisms that use sex are always part of a particular kind of group – in which genes are shared among its members. We call this group a species. While it is true that asexual organisms also form groups called species, they do not share genes. Species are, by definition, members of a group that exchange genes with one another. Actually many species definitions exist (these will be discussed later), but here I am interested only in those species that reproduce sexually, or had sexual ancestors.

The 1999 papers by the Kondrashovs' and by Dieckmann and Doebeli had one goal – to explain how new species could evolve in an interbreeding population. That is, they were attempting to explain the evolution of *sexual species*. Speciation, in their model, occurs because a population diverges in mating habits as it undergoes adaptation to an available niche. While their models were concerned with speciation, they may also explain why sex exists. This was clearly not the intention of these scientists and they would probably argue that their theories do not accomplish this goal. I believe that they do, however, and will devote much of the book to advancing this idea.

Adaptive speciation produces organisms able to take better advantage of their environment. Because sex is essential to this form of adaptation, *every adaptive speciation event is an act of selection for sex*. If the advantages of speciation are high enough, it might overcome the many disadvantages of sex. Once speciation is complete, sex may provide no additional advantage over asexual reproduction, but conversion to the more-fit asexual mode would not be a simple matter, and the new species will remain sexual indefinitely.

This last fact is critical. For many sexual species, including all birds and mammals, no viable pathway exists that could produce an asexual variety. Even among groups where asexuals exist, they are rare and all of these, as far as we know, arose from a sexual ancestor. To understand sex, we may need to ignore the birds and the bees and focus on the yeast and the sponges. Indeed, we may need to go even further back in time to when the first single-celled sexual organisms were evolving.

We should also consider the full genetic consequences of sex. Genetic variation, which has been the focus of most ideas about sex, is only half of the equation. Sexual organisms, all of them, limit genetic variation through reproductive choice. Birds of a feather not only flock together, thy mate together. Again, we can treat this fact in two logical ways. Individuals choose a mate *because* he or she is of the same species, or individuals are members of the same species *because* of how they chose a mate.

Reproductive choice is an essential element of speciation, and one aspect of reproductive choice, sexual selection, has long been recognized as a promoter of speciation. Sexual selection operates on traits that seem to have no adaptive value, but only serve to attract a mate. The feathers of the peacock are probably the best know example of sexual selection. In 2001 a group from Holland, led by G. Sander Van Doorn, reported on the results of extensive analysis of sex-related genes and their variation among different species. Their results led them to conclude that: "Sexual selection . . . and sympatric speciation are interwoven processes . . ." Said another way, sexual selection leads to sympatric speciation, while sympatric speciation promotes sexual selection.

This mechanism of speciation may seem to have limited application. It seems, for example, that it could not apply to a plant. Can an orchid select its mate? It certainly tries. The shape, color, odor and timing of its bloom all help ensure that its pollen will be carried to a compatible partner, one of its own species. For a flowering plant, the brain of a bee or a butterfly substitutes for its own lack of a nervous system. In fact, a brain is not a requirement for reproductive selectivity. Many animals lack eyes or a nervous system. Yet, in its own way, a starfish egg is as particular about its mate as is a peahen is about hers. The proteins that line the surface of a starfish egg screen the sperm that collide with it, and allow only the most "beautiful" one to gain entry.

Later I will discuss the controversial issue of how sex first evolved, but no matter how it happened it had an important consequence – sex enabled more rapid adaptive speciation. About 550 million years ago, at the start of the Cambrian period, life was simple. Fossils of multicellular life forms are nearly absent from this period. Within a few million years, however, an explosion (the event is called the Cambrian explosion) of animal types occurred. Various theories have been proposed for the Cambrian explosion, but like most ancient evolutionary events we will probably never understand it fully. However, it almost certainly involved sex. Mary Droser, of the University of California, Riverside, had this to say about fossils of the tubular animal *Funisia dorthea*, which date to about 570 million years ago: "In *Funisia*, we are very likely seeing sexual reproduction in Earth's early ecosystem – possibly the very first instance of sexual reproduction in animals on our planet."

While *Funisia* is clearly an early form of a sexual animal, it was by no means the first sexual organism on our planet. All sexual organisms, plants, animals and fungi, use the same basic sexual processes. This tells us that sex began with a single-celled organism that was the ancestor to the more complex forms that were to come.

It has been proposed that the Cambrian explosion was driven by the evolution of the eye. The eye helps animals find food and avoid predators, and was unquestionably a major evolutionary invention. It certainly would have been important way to increase fitness, but the eye may have had another role in increasing diversity -- it provided a new means of selecting a potential mate, and a new pathway to speciation. Whatever the cause, the Cambrian explosion would have been the Cambrian fizzle if it were not for sex.

## Sex and speciation: One topic or two?

In his popular book on the evolution of sex, *The Red Queen*, Mark Ridley presents a fable about Martians, who are assumed to be asexual organisms. As told by Ridley, the sexless Martians attempt, without success, to understand the prevalence of sex on planet Earth. They are astounded that such an inefficient reproductive mechanism should be so common, and surprised that human biologists have been unable to explain this fact. It is an amusing approach, but it seemed to me that the premise is unbelievable. Not that Martians per se are unbelievable. What is unbelievable is that the advanced, sentient beings of Mars could be asexual.

Could asexual organisms evolve into human-like beings? We will never know for sure, but the experiment has been done here on Earth, where asexual organisms, such as bacteria, abound. Indeed, they make up, by far, the majority of living organisms. Yet these organisms have never found a way to get bigger or more complex than single cells. Many explanations could be given for this remarkable fact, but based upon this single data point purely asexual organisms seem unlikely to evolve into complex forms. Fantasies about sentient asexual organisms on Mars may be amusing, but are only fantasies. If asexuals were capable of evolving into complex organisms, they had ample opportunity to do so on Earth, but did not.

Biologists have largely ignored the idea that sex exists because of its role in speciation, and the few who have proposed such a connection have been mostly ignored. Indeed, if the widely accepted theory of allopatric speciation is valid, then sex should actually be an impediment to speciation! Geographical isolation is invoked in allopatric theories precisely because it prevents the mixing of genes that are a result of sex. Asexual organisms, in contrast, have no need of geography to prevent gene flow, so asexual species should exist in abundance. On the contrary, all of the advanced, multicellular species that occupy our planet arose from sexual ancestors. None are a product of a purely asexual line of descent.

The ideas presented here are sure to be controversial in some quarters. As I develop them I will address, as best I can, the most serious objections. For the moment here, briefly, are some of the issues that I will consider:

While I focus on speciation, full speciation is the endpoint of assortive mating, the nonrandom association of reproductive partners. Assortive mating increases fitness by combining favorable parental genes in the offspring, and helping to preserve such genetic combinations through subsequent generations.

Population size is important. The events that lead to speciation are rare, per reproductive cycle, but become more likely as population size increases. Evolution typically depends upon rare events, so arguments that a particular process is "unlikely" are not an argument against its importance. A few individuals in a population that are simultaneously more fit, and prefer to mate with partners that have these same genes, can have a large, rapid evolutionary impact.

A good deal of effort has gone into attempting to explain how sexual populations can survive when they are in direct competition with asexual relatives. We should not get too distracted by this issue. The asexuals arose from sexually reproducing ancestors. All of them. The real question is not why the asexuals don't take over, but why sexual species occasionally produce asexual progeny.

A major argument against the proposed role of speciation in the maintenance of sex is that speciation is too slow to have a significant impact. Indeed, this is probably true – if sex imposes a twofold reproductive cost associated with the "cost of males." When sex first evolved, however, males and females did not exist. Sexual recombination would have occurred between equal partners, and the reproductive costs to either parent would have been minimal. The potential benefits of assortive mating, and speciation, may have been more than adequate to overcome the small costs of genderless sex. Males evolved millions of years after sex became commonplace, and by then sexual species had become the dominant life form.

Adaptive speciation enhances fitness. As a consequence, traits that promote adaptive speciation will be selected for, and may be more common than generally believed. For example, chromosomal linkage between reproductive and ecological genes is known to favor speciation, but is generally thought to be rare. On the contrary, such linkages may be commonplace, precisely because they promote speciation.

Trying to explain sex based upon modern organisms may prove impossible. Sex first appeared in a world long gone, and we may have to use our imaginations to understand that world. Once sex was established, along with its ability to promote rapid adaptive speciation, there would have been no going back.

Our understanding of species and speciation is undergoing a revolution through the use of DNA analysis. These studies reveal that many more species exist than previously thought, and that new species can emerge rapidly under some conditions. Sympatric speciation, once considered rare, is becoming more accepted as both possible and common. The role of sex in speciation cannot be denied, but the impact of this fact on our understanding of sex has, so far, been minimal.

I will have much more to say about these issues in the remainder of the book. I will not, however, prove the validity of any particular theory. This is simply not possible. Not for my ideas, nor any of 40 or so theories of sex that have been proposed. Given the power of evolutionary theory, it may seem odd that such huge gaps should exist in our understanding of issues as important as speciation and sexual reproduction. Behind this failure, I suspect, is a fundamental limitation of evolutionary science.

## Physics envy: The limitations of biological theory

A physicist asked to explain the structure of matter will tell you about the "standard model," which explains quite nicely the properties of atoms. Other essential aspects of the physical world, such as gravity and the nature of light, have their explanations in Einstein's General Theory of Relativity and Quantum Theory, respectively. These theories explain, to most physicists' satisfaction, the behavior of the universe.

Biologists have their theories too. Evolution being the main one. As a rule, Darwin's theory has proven as powerful in biology as those of Newton and Einstein in physics. Yet, cleaning up remains to be done, and the issues of sex and speciation stand out because of their importance. The scientists who study these issues are certainly smart, and many are brilliant and creative. So why have these issues remained unsolved?

I suspect that the theorists have become trapped in their equations. Darwin did not rely on mathematics for his analysis. He was unaware of Mendel's work on the rules of inheritance, but was still able to make a solid case for evolution by natural selection.    Today, however, it is impossible to imagine evolutionary biology without the mathematics of genes.

Yet, equations are only as powerful as the data that goes into them. A bitter truth about evolutionary science is that hard data is extraordinarily difficult to come by. Field observations of the selective forces acting on a population are always difficult, if not impossible.    Organisms can die of many causes. Predators, diseases, starvation, toxins and accidental injuries all take their toll. The survivors of these assaults must still find a mate and raise offspring if they are to have an impact on evolution. Assuming the necessary data on mortality and reproductive success can be obtained, we would still need to know about the genes involved to predict evolutionary consequences. It is not enough to know that some rabbits get eaten by foxes and others by hawks, and some avoid both of these predators, only to die of starvation or disease. To make evolutionary sense of these differences we would need to know if they have a genetic basis.

Under ideal conditions, with simple organisms, living in uncomplicated environments, the selective pressures and their genetic consequences can be analyzed. For most species, however, life is too complicated, and evolution is too slow. Evolution is a historical process, and often the best we can do is imagine the past in an attempt to explain the present. We do this even though key events often took place well before biologists, or humans of any kind, were around to observe them.

As an example, consider the horned lizard of the American Southwest. The formidable row of spikes that give it its name look intimidating, and they appear to be defensive weapons. But against what? In my youth, growing up in San Diego, I used to catch them frequently, and the horns were no obstacle at all. A team of biologists from Utah State University noticed that the heads of these lizards could frequently be observed hanging from branches. Just the heads, hanging by their horns. These were the dinner remains left by shrikes, an avian predator of the horned lizard. When a shrike makes a catch, it hangs the lizard by its horns on a nearby branch, consumes the body and leaves the head behind. The horns of the decapitated lizards were measured and compared to that of living horned lizards. The shrike's victims had, on average, shorter horns.

The Utah biologists concluded that the horns evolved because they protected the lizards from shrike attacks. This perfectly reasonable conclusion was roundly criticized. Others pointed out that the horns may have evolved in response to some other predator in the distant past, or perhaps the horns are sexual attractants, like the peacock's feathers, or used for intraspecies combat, like the ram's horns. Even if the conclusion is correct, it does not explain why other lizards, which are also attacked by shrikes, lack horns. Indeed, shrikes may prefer horned lizards as prey precisely because of how easy they are to hang by their heads! More extensive observations would help resolve these issues, but in truth we are unlikely to ever fully understand the evolutionary history nor the present selective forces on horned lizards.

Biology is replete with unresolved issues of this sort. The black mask on a raccoon's face, or the snowy underside of a white-tailed deer's tail, are fascinating subjects for speculation, but remain a mystery. If biologists cannot agree why a horned lizard has horns, or a white-tailed deer a white tail, then what hope do we have of understanding something as complex as sex? In truth, we may never reach a consensus explanation for sex, but this won't stop us from trying.

Recently, a study of rotifers (microscopic animals characterized by a mouth lined with beating cilia), found that a heterogeneous environment favored sex in a species that could switch between sexual and asexual reproduction. This result supports a theory that sex evolved because it increases fitness in a varying habitat. Other evolutionary biologists when asked about the results had a more cautious response. From Brian Charlesworth, of the University of Edinburgh, "It is suggestive, but I don't think people are going to go around saying 'Ah, we now know why sex evolved." And from Rufus Johnston of the University of Cambridge, "This could just reflect rotifer breeding behavior rather than telling us anything about the evolution of sex."

So even when correspondence is demonstrated between theory and experiment, a final verdict remains elusive. This reflects the harsh reality of evolutionary biology. Which is not to say that we must give up all hope. Each theory, each experiment, adds to our knowledge, and becomes an element of future discussion.

I once saw an article on speciation in which the relevant equations included 20 adjustable variables. Good luck on confirming the validity of these equations in the real world. I have the greatest admiration for the model builders. Their computers and equations have greatly improved our insight into the mechanics of evolution. Yet for many situations the inputs to these equations are mostly guesswork. Verbal theories in evolution, as opposed to mathematical ones, have taken a beating in recent years. This is unfortunate, as even the most sophisticated equation begins with an idea. Perhaps the topics of sex and speciation are in need of new ideas, from which new quantitative theories may ultimately emerge.

I am not the first one to suggest a linkage between sex and speciation. This view appears to have first been raised in 1975 by S. M. Stanley, whose article "Clades versus clones in evolution: Why we have sex." proposed such a link. He was followed in 1989 by Leonard Nunney, whose article "The maintenance of sex by group selection" proposed a modified version of Stanley's hypotheses.

I will discuss these ideas in detail later, but what is striking is that these papers have been nearly forgotten. A 2010 review on the evolution of sex by Stephanie Meirmans and Roger Strand has more than 70 references, and no mention of Nunny or Stanley. While a 2009 review by Sarah Otto has over 90 references, and also manages to omit the papers by Nunney and Stanley. A 2008 review by Lilach Hadany and Josep Comeron does site Nunney, but not until the last paragraph, and only on a peripheral issue.

In the age of the internet, these articles cannot have been simply lost (I found them with a simple Google Scholar search). For reasons that are not clear, they have been consigned to scientific oblivion. They don't get slammed with devastating criticism, they simply go unmentioned. I have struggled to understand why these two have been subjected to a kind of scientific silent treatment. Did they lack the credentials to be taken seriously? No. They were both respected professors at reputable universities. Did they publish in obscure locations? No. Stanley's paper was in Science, one of the most respected journals, while Nunney published in Evolution, the official organ of the Society for the Study of Evolution.

I think I know why these articles have been ignored. Both gave speciation a central role in explaining the prevalence of sex. This ran counter to the widely held view that speciation is largely an accidental event caused by geographical isolation. Nunney's proposal that group selection on species is a primary reason for sex was simply too much for most biologists to swallow. Not only did he invoke group selection, a concept that has been discarded by many biologists, he suggested that species compete with each other for resources. This made no sense if species are assumed to arise in isolation.

It is time, I believe, for biologists to take a second look at these ideas.

# Chapter 3: Biology 101 and beyond

## Term limits – genes, alleles and evolution

This is not a textbook. Nevertheless, certain topics will require a basic understanding of biology.  In particular, three terms – genes, alleles, and evolution – need to be clarified before we can explore sex and speciation adequately. Let's start with the concept of gene.

A gene is a part of DNA that controls one or more traits of an organism. Most genes provide the instructions to make a protein, and it is the protein that performs the actual work of the gene. The mere existence of a gene, however, does not mean that the protein specified by that gene will be made. Genes can be turned on and off, and the control of gene activity is as important as the genes themselves.  Humans, for example, have a gene that is responsible for the formation of sensory whiskers and "penis spines." (Penis spines are fleshy projections found on many mammalian penises, which probably act to extract vaginal semen deposited by other males.)  We lack such projections on our faces and penises not because we lack the gene, but because of a DNA deletion that removed a regulatory site that controls the expression of this gene. (Why this deletion spread through the evolving human species is unknown.)

Humans, according to recent estimates, have about 25,000 protein coding genes, and an unknown number of regulatory ones. These genes are distributed among 23 chromosomes, each of which consists of a molecule of DNA. A person has two copies of each chromosome, one from each parent. The exception is the Y chromosome, which is only found in males. The number of chromosomes in different species varies greatly, even among closely related organisms. For example the Indian muntjac, a small deer, has just three kinds of chromosomes, the smallest number known in any mammal. Yet a close relative from China has 23 chromosomes.  The muntjac is an extreme example, but as a rule closely related species will differ in chromosome number or architecture. Later I will suggest an explanation for this odd fact.

Genes carry genetic information, and we can use other information carriers as reasonable metaphors for genes. For example, consider the English sentence:

*Eat beans.*

These words, like a gene, provide instructions for a course of action. Now consider the sentence:

*Make haste.*

These words provide a different set of instructions and represent a different "gene." Evolution depends upon variations in genes, which take place by mutation. We can "mutate" our sentence "gene" by changing one or more letters. For example, replacing the *n* with a *d* in the first sentence leads to:

*Eat beads.*

We can also mutate the second sentence, for example, to read:

*Make waste.*

Biologists use the term **allele** to describe different forms of a single gene. Thus, *Eat beans* and *Eat beads* are alleles of the "Eat" gene, while Make *haste* and *Make waste* are alleles of the "Make" gene. Mutations and alleles are closely related, as alleles are a result of mutations. (The main difference between the terms, as generally used, is that alleles are established variations in a species' DNA, while mutations are recent alterations.)

A sexually reproducing organism will, as a rule, have two copies of the chromosome that carries a particular gene (one from each parent), and these can exist in only three possible combinations, or **genotypes**. For example, the *Eat beans* gene and its allele, *Eat beads*, can form the following genotypes:

Genotype 1: *Eat beans/Eat beans*
Genotype 2:. *Eat beads/Eat beans*
Genotype 3: *Eat beads/Eat beads*

When the two copies of this gene are the same, as in genotypes 1 and 3 (called the *homozygous* state), the instructions are clear. You either eat beans or beads. When they differ as in genotype 2 (called the *heterozygous* state), you would seem to have a choice. Do you eat both beads and beans, or does one set of instructions take precedence?  A similar issue arises with the "Make" gene. What would be the consequence of the genotype *Make haste/Make waste?* To answer this question we need to deal with the concept of genetic dominance.

## Dominance and the fate of the twin-engine airplane

Today scientists have a good understanding of the basis of genetic dominance. Yet the concept of dominance gives many people problems, in part because the term itself has certain connotations in English that do not reflect its scientific meaning. The following story should help clarify the basis of genetic dominance:

Five biologists are flying to a conference in a two-engine airplane. Suddenly, one of the engines makes an unpleasant noise and starts emitting smoke.

"I hope that engine doesn't quit," says the first biologist. "A bad engine is probably dominant and we will crash for sure."

"I wouldn't worry," says the second biologist. "A bad engine is most likely recessive, and the airplane will fly normally on just one working engine."

"I am not so sure," says the third biologist. "I suspect the neither the good nor the bad engine is dominant. They are co-dominant and the airplane will still be able to fly, but at only half its normal speed. We will reach our destination, but we will be late for the conference."

"You guys should get a life!" says the fourth biologist. (We'll hear from the fifth one in a moment.)

Like a twin-engine airplane, the functionality of most sexually reproducing organisms depends upon the activity of pairs of genes, the unavoidable consequence of having two parents (An exception to this rule may occur with genes located on a sex chromosome). In any individual, the two copies of a gene may be the same, or they may differ. As a result, as I mentioned previously, a single gene can be present in only one of three possible combinations, or genotypes.

Let's designate the engine "gene" of our hypothetical airplane by an E, and use the standard notation of uppercase for the dominant form, and lowercase for the recessive form. The airplane can have only three possible states, designated as EE, Ee, and ee. Clearly, if both engines are good the plane flies normally, and if both are bad, the plane crashes. When discussing dominance, we are only concerned with the state in which the two forms of a gene differ, the heterozygous state, the Ee combination in my example.

The argument between our biologists is whether to use an uppercase E for the good engine, or for the bad one. Most likely their airplane will fly more-or-less normally on one engine. A working engine, in genetic terms, is probably "dominant" over a malfunctioning one. Indeed, one reason many airplanes come with two engines is to provide a margin of safety should one engine fail. In a similar way, having two copies of a gene can protect an organism if one copy should go bad. For many genes, a single good copy is sufficient to maintain the function of that gene, and the other copy is present merely as backup. In such cases, an organism will display only the trait determined by the dominant form, even if it carries a copy of the recessive gene.

Of course, things can be more complicated. Perhaps the third biologist is right, and the airplane will continue to fly on one engine, but at a slower speed. In this case, the uppercase-lowercase symbolism is inadequate. Biologists call such situations *codominance* or *incomplete dominance*, and will typically use superscripts to designate the different forms of a gene. Thus we might indicate a good engine by $E^g$ and the bad one by $E^b$, and the "genotype" $E^g E^b$ will still fly, but at an intermediate speed.

The fifth biologist in the airplane, whose specialty is evolution, also has an opinion: "You are all wrong. We cannot consider the engines in isolation. Whether or not the plane stays in the air will depend upon the environment. If we hit a storm we will crash for sure, otherwise we should be ok. Furthermore, an airplane requires more than engines to fly. If some other component should fail, one working engine may not be enough to keep us airborne."

I hope the two-engine airplane metaphor makes the concept of genetic dominance clear. If not, that's ok. Genetic dominance is probably of little importance in understanding sex or speciation. It may have actually led scientists astray.

Let me explain.

Like many teachers of biology, I found that my students struggled to understand genetic dominance. For years I wondered how such a simple concept could be so difficult. I think I now understand my student's struggles – genetic dominance contradicts what they know about inheritance before they even enter my classroom. The young people in my classes were neither ignorant nor stupid. They were well aware of certain facts of genetics. They knew, for example, that if two people of different skin pigmentation married, their children would most likely have an intermediate skin tone. Indeed, as a general rule, children tend to look like a blend of their parents. This knowledge did not help my students in Biology 101, with its focus on simplified genetics.

Most biology textbooks introduce the concept of dominance with the example of Mendel and his pea plants. Part of Mendel's genius was that he restricted his studies to traits that allowed quantitative analysis. Most of my students, however, have never seen a pea flower. They have seen plenty of other flowers, but they rarely resemble the pea flowers in their textbook. Flowers come in many colors and shades, not just purple and white, and they often display complex patterns of different colors. Good students easily memorize the rules of Mendelian genetics, but in the back of their mind they must think that these rules only apply to pea plants.

Many textbooks also provide examples of dominant and recessive traits in people. Here is a list of the more common ones:

Tongue rolling (dominant)
Free hanging earlobes (dominant)
Cleft chin (dominant)
Hitchhiker's thumb (recessive)

Dimples (dominant)

Knuckle hair (dominant)

Certain features are interesting about this list. First of all, none of these have any obvious relationship to fitness. We presumably do just fine whether our ear lobes hang free or are attached to our skull, or whether we have hair on our knuckles. I suppose that curling one's tongue into a tube might be beneficial in some situations, although I can't think of any off hand.

A second feature of this list is that none of these genetic traits are likely the reason for the gene in question. Humans, as I indicated previously, have about 25,000 genes. Are we to believe that one of these genes is devoted to the exclusive task of controlling the hair on our knuckles? More likely, knuckle hair is a side effect of a gene that has other, more important, functions. We just don't know what those other functions are.

It is no wonder my students are so confused! We give them examples that contradict their actual experience of how genes work, or which relate to trivial side effects of unknown genes.

But I am supposed to be discussing sex and speciation, not the inadequacies of introductory biology textbooks. Have patience. As I get to the details, we will discover that neither sex nor speciation can be properly understood if we confine ourselves to simple genetics. We need something more than pea flowers or knuckle hair as examples if we are to understand the role of genes in sex and speciation.

Consider human evolution. Which of the following might have been important to the survival of our ancestors a few hundred thousand years ago?

a, Intelligence

b, Running speed

c, Skin pigmentation

d, Ability to speak

e, All of the above

I assume that you were able to pick as the correct answer e, all of the above, from this list. What is notable is that these traits are all genetically complex, and each is controlled by several genes. While you might find references in the popular media to the discovery of a "gene for intelligence" biologists understand that brain function depends upon a great number of genes working in concert.

Traits, like skin color and body height, that are controlled by several genes are called "quantitative" traits. A quantitative trait varies, more-or-less, in a smooth fashion over its range. Unlike Mendel's pea flowers, which were either purple or white and nothing in between, humans come in many different colors. We also come in many different heights and IQ's. During sexual reproduction the various genes that control quantitative traits are randomly distributed to the offspring. As a consequence, when multiple genes are involved, offspring tend to look like a blend of their parents.

Individual genes are important, and some mutations are, by themselves, beneficial or harmful. However, explaining complex phenomena like sex and speciation requires more complex genetics than that discovered by Mendel.


## Linkage: When genes travel together

As a rule, genes are linked if they are located on the same chromosome. Humans have 23 pairs of chromosomes and about 25,000 genes. Simple math tells us that each chromosome must contain on average about a thousand genes. Each of these genes is linked physically to all of the others for the simple reason that they are attached to the same molecule of DNA.

Consider our example of sentences. The *Eat* gene and the *Make* gene might be linked, which we could illustrate as:

*Eat beans.- Make haste.*

Or perhaps they are unlinked:

*Eat beans.*

*Make haste.*

Linkage has a profound impact on evolution. For example, suppose that a gene undergoes a beneficial mutation. Nearby, on the same chromosome, are many other genes. As the beneficial mutation spreads through the population, nearby genes get a free ride and share in any success created by the good mutation. Of course, even good genes can lose out if they sit too close to a harmful mutation. Geneticists refer to this phenomenon as *hitchhiking*, and like a hitchhiker a gene goes where its ride goes. For better or worse.

Linkage can be broken by a phenomenon called *crossing over*. Let's try a metaphor: Imagine two sisters, call them Alice and Betty, who are dressed for the big dance. Alice and Betty represent a chromosomal pair, and their clothing, consisting of hat, blouse, skirt and shoes, represent genes. These "genes" are *linked* because they remain together when Alice and Betty go to the dance. Although they are sisters, they make different fashion statements. Alice wears a blue hat, green blouse and white skirt, while Betty sports a purple hat, pink blouse and red skirt. They both wear ruby slippers, which we will say no more about. The different colors, in this metaphor, represent different versions (alleles) of the hat, blouse and skirt genes.

At the dance, the girls decide that their outfits could use some improvement. They slip into the lady's room and swap blouses. As a result, they each make a new fashion statement. Alice now wears a pink blouse to go with her blue hat and white skirt, while Betty now wears purple, green and red. Both girls are still fully dressed, and all that has changed is the combination of colors. Has there been any improvement in their fashion sense? That's hard to say, given the ever-changing rules of fashion.

Consider again the sentence *Eat beans - Make haste*. This combination makes a certain sense (Eat your beans, and be quick about it!), at least when compared to *Eat beads - Make haste.* The combination *Eat beads - Make waste* also makes some sense (If you eat beads, you want to be turning them into waste!). Genes never work alone.  What is important is that they function harmoniously as a unit.

From an evolutionary perspective, the important question is whether a new combination of alleles enhances fitness. This question has no simple answer, as fitness, like the rules of fashion, depends upon the local environment. Swapping genes may increase or decrease the fitness of one's offspring, but only time will tell which it is.

This is enough basic genetics for now. For the average person this lesson was probably unnecessary. Unless your mind has been corrupted by too much academic biology, you probably already understood what is important:

1. Complex traits are controlled by several genes, which appear to blend the parental types.

2. Genes never act alone, and must function harmoniously for optimum fitness.

Ultimately, the fitness of an organism depends upon how its genes interact with the local environment. These interactions will determine the course of evolution, which I discuss next.


## A primer on evolution

Let's get one thing straight: Evolution is a fact. All things change over time. Rivers carve out valleys. Stars burn out. Mountains erode. These are all examples of evolution. The genetic composition of living organisms also changes from generation to generation. Your children will look different from you, just as they will look different from their children. This is biological evolution.

Of course, when biologists use the term evolution they usually have a particular type of change in mind, as first envisioned by Charles Darwin. This is the principle of evolution by natural selection, also called the *theory* of evolution by natural selection.

Much has been written about the term *theory* and how it is used in respect to evolution. As commonly used, the word "theory" implies an idea that is not yet proven. Biology textbooks, however, will inform you that a theory in science is a widely accepted principle of nature, and that just because Darwinian evolution is called a theory does not mean that it is unproven.

The terminology controversy is pointless. Newton's rules of force are called laws, but Einstein's principle of relativity is called a theory. In biology we have the cell theory, but no DNA theory. Mendel's discoveries about inheritance may be called a principle, rule or law, but never a theory – go figure! Evolution by natural selection is a fact, and whether we call it a theory or a principle or a law does not change that fact. Just because a few people do not believe in evolution does not turn it into an unproven theory.

Some people argue that evolution by natural selection could, in principle, be proved wrong, and thus should be called an unproven theory. This is utter nonsense. No conceivable observation could overthrow the principles of evolution. Creationists, of course, provide a host of "facts" that disprove evolution. They are like the conspiracy theorists who argue that the Apollo moon landings were faked. We should pay no attention to them.

For amusement, let's assume that the creationists are right, and that our world is the invention of an all-powerful creator. This would not erase the extensive body of evidence in support of evolution. Suppose that God should reveal Herself, and announce that She had created the Earth and all of its life forms a mere 7,000 years ago, and that this only took Her a week! This would be a shocker, but scientists would still have to deal with the reality on the ground. The Creator's confession would not alter the fossil record, nor DNA molecules, nor suddenly make the equations of evolutionary theory invalid. Facts, as the saying goes, are facts, and evolution by natural selection is one of these.

I have long felt that scientists have done a lousy job of defending evolution. They need to be reminded that the best defense may be a good offense. When I have had occasion to debate creationists I let them have it: Where is your data? Give me some testable predictions. If a Creator made everything – who made the Creator? Can you write a computer program that incorporates what we know about life, and which doesn't lead to evolution? Why do you accept some parts of the bible, such as Genesis, to be literally true, and not other parts, such as the locusts "with breastplates of iron" in Revelations?

I understand why many people have trouble with evolution. They have difficulty understanding how complex organisms could have emerged from the randomness of mutation. While mutations may be random, the process by which favorable ones get selected is not. Favorable mutations get passed on, but unfavorable ones do not. It is that simple.

The other problem many people have is that they have a poor sense of large numbers. Life has existed on Earth for about 3,500,000,000 years. When I teach this topic, I ask students to imagine that life as a 350 page book, in which each page represents 10,000,000 years. On this scale, humans do not appear until we are in final paragraph on the last page. All of recorded human history is on the last line, and a human lifetime is represented by a few ink molecules in the period at the end of that line.

The unit of time in evolution is not years, however, it is generations. For this reason I get a bit upset over references to "simple" single-celled organisms. The microorganisms that occupy a drop of pond water are not simple. At a cellular level they are more complex than any of the cells that make up your body. This is understandable when we realize that they have been subject to many more generations of evolution than the "higher" plants and animals of this planet.

Evolution is like a lottery. The odds of winning are very small, but eventually somebody wins. Keep these facts in mind when I discuss the early phases of sexual evolution. Highly unlikely events must have occurred, but with trillions of cells competing for existence over trillions of generations they were going to happen eventually.

### Preadaptations and exaptations

I write these words using bits of anatomy that evolved for quite different purposes. My opposable thumbs, originally used for gripping objects such as rocks and clubs, now fit nicely over the spacebar on my keyboard. Finding new uses for existing structures is not unique to humans, and is one of the essential principles of evolution by natural selection.

Feather-like structures, for example, first appeared on flightless dinosaurs. These probably provided thermal insulation and may have also been used for sexual displays. They had nothing to do with flight, but nevertheless were an essential foundation for the eventual evolution of bird flight. Traits that evolve new functions are called preadaptations (biologist Stephen J. Gould preferred the term exaptations, but this does not appear to have caught on). For example, the air bladder of fish was a preadaption that made possible the evolution of the lungs of reptiles and mammals.

Preadaptations lead to new adaptations, which become preadaptations for newer adaptations. And so on. Sex, to stay on topic, evolved out of something other than sex. Later, I will suggest a preadaptation that may have led to sex, and how it may help us understand the role of sex today.

Furthermore, sex may itself have served as a preadaption for something else important, namely adaptive speciation.

## An imperfect world

I love nature shows on television (particularly since I got HDTV). I do not, however, care much for the way they sometimes butcher the science of evolution. Tops on my list of irritations is the phrase "perfectly adapted." Organisms are never "perfect." Perfection, apart from the scoring in some sporting events, does not exist. I am not even sure how it can be defined. Evolution is a competition, and unlike figure skating there are no perfect scores. To succeed in nature, one does not need to be perfect, only better than the competition. But no matter how successful you may be today, tomorrow you may be a loser.

It is understandable, however, that many people believe that evolution leads to perfection. With each generation, the most successful individuals reproduce, leading to the gradual spread of their genes. Logically, a few million cycles of reproduction and selection would leave little room for any further genetic improvements. Each species should be as perfect as possible given the amount of time they have had available for their evolution. If this is so, however, why does so much genetic variation exist within a species? Biologists have struggled to answer this question, but have not reached a consensus. I will not discuss these explanations here, which are not crucial to the discussion of sex and speciation. What is interesting, in the present context, is the relationship between sex and genetic variation.    Three facts about genetic variation are worth emphasizing:

Genetic variation is essential to evolution.

Genetic variation is increased by the mixing of parental genes that occurs as a result of sexual reproduction.

Genetic variation produced by sex is limited by the mate choice mechanisms of speciation.

Theories of sex focus on the genetic variation it produces, but sexual organisms are not interested in sharing genes randomly. They are selective, and remarkably so, in their choice of partners. These choices are designed (by evolution) to restrict the genetic variation of their offspring. If you do not believe this statement, then take a deep breath – and be thankful that the assortment of male sex cells (pollen) that you inhale are likely to produce nothing more than a bout of hay fever. Certainly no goldenrod-human offspring.

Ok, this is a bit farfetched, but consider that goldenrod pollen will only fuse productively with goldenrod eggs. For the goldenrod, your respiratory tract is a dead end and a waste of reproductive potential.  All around us organisms are having sex, but they do so, almost exclusively, with their own kind. Biologists call these reproductive kinds "species."

Each species is characterized by a set of genes that work together in harmony, and sexual reproduction must maintain these harmonious gene combinations, even as it creates new ones. This is the true paradox of sex.  Somehow sexual reproduction both creates and protects biological species. To solve the mystery of sex we may also need to solve the mystery of speciation.

# Chapter 4: What is a species?

## The first chicken

You may own one of the many guidebooks that identify the birds, butterflies, trees and wild flowers that grace the landscape. Such books enable you to distinguish, for example, a bluebird from a blue jay. It will not, however, tell you why these birds exist as distinct species. Darwin, as I noted, was among the first to ponder this issue, and asked "Why are not all organic beings blended together in an inextricable chaos? Instead of chaos, living organisms can be assigned to distinct types, its species, and we do not fully understand why or how this happens. To begin, we need to consider what, precisely, a species is. I'll start by considering another classic question: Which came first the chicken or the egg?

The chicken or the egg question seems to be an insoluble problem. After all, you need a chicken to make an egg, but you only get a chicken from an egg to begin with. Nevertheless, there is a correct answer. The egg came first.

Let me explain. A chicken belongs to a particular species, *Gallus gallus*. (This is also the name of the red jungle fowl, the ancestor of the domestic chicken, and the following discussion is really about this wild bird.) While many varieties of chickens exist, they are identifiable as chickens because they have a particular set of genes (and their alleles), and the question is whether the chicken or the egg had these genes first.

Many years ago, before chickens existed, two chicken-like birds mated. The female subsequently laid an egg that contained, for the first time ever, chicken genes. The egg hatched, and out walked the Earth's first chicken. Just like that! So what happened to the idea that evolution is a slow, gradual process?

Consider that very first chicken again. It will not only be the first chicken on the planet, it will be the only one. There will be no chicks in this chicken's future. We can assume, however, that the first chicken was a survivor and that its offspring were also survivors. We can safely assume this because chickens are with us today. Over time, chicken genes spread through the evolving population, and with each generation the odds of a chicken being born increased, as did the chances of two birds with chicken genes mating. Eventually, these genes spread to all of birds in this population. A biologist would find distinct differences between these birds and others of a similar type, and she might even be moved to declare them worthy of their own unique species designation – *Gallus gallus*.

Which is a problem. Biologists do not agree on what a species is. For example, John S. Wilkins, of the University of Sydney, has compiled a list of 26 species concepts, which clearly complicates things if one wants to discuss how species evolve. Charles Darwin had a cynical view of the issue. In a letter to botanist Joseph Dalton Hooker he wrote: "It is really laughable to see what different ideas are prominent in various naturalists' minds, when they speak of 'species.'...It all comes, I believe, from trying to define the undefinable."

Nevertheless, define we must! Below are some species definitions, considerably fewer than 26, beginning with one, sort of, from Darwin, followed by definitions (modified from a Wikopedia list) that look at species from different perspectives:

## Species definitions

Darwin's view:  "I look at the term species as one arbitrarily given for the sake of convenience to a set of individuals closely resembling each other ... it does not essentially differ from the term variety, which is given to less distinct and more fluxtuating forms. The term variety, again in comparison with mere individual difference, is also applied arbitrarily, and for mere convenience sake."

Biological species concept (BSC):  A set of actually or potentially interbreeding populations. (This is the most widely adopted species definition.)

Typological species:   A group of organisms in which individuals conform to certain fixed properties.

Morphological species:    A population or group of populations that differs morphologically from other populations.

Reproductive species:  Organisms that are able to reproduce naturally to produce fertile offspring.

Mate-recognition species:    A group of organisms that recognize one another as potential mates.

Phylogenetic species:  A group of organisms that shares an ancestor; and are part of a lineage that maintains its integrity with respect to other lineages through both time and space.

Ecological species:   A group of organisms adapted to a particular niche within their ecosystem.

Genetic species:   A group of organisms having an agreed upon similarity of DNA.

Cohesion species: "Most inclusive population of individuals having the potential for phenotypic cohesion through intrinsic cohesion mechanisms."

Evolutionarily Significant Unit (ESU):   An evolutionarily significant unit is a population of organisms that is considered distinct for purposes of conservation. This is not, strictly speaking, a definition of species, but it serves as a stand-in for determining when an organism is worthy of legal protection.

What a mess! We should not be totally surprised by this state of disorder, however, as nature does not act for the convenience of biologists. Species do not fit into a neat, definable package.   Nor could they, since species are a product of evolution, and what we see today is a snapshot of a historical process.  Over time some species will disappear, and others will emerge from what are now single species.  Some organisms even have the nerve to mate outside of their own species, ignoring the biologist's requirement for reproductive purity.

## The species problem

Scientists and philosophers have fussed over the definition, and even the reality, of species for years, but without resolution. They even have a term for their dilemma. Uncreatively, they call it the "Species Problem."  Most biologists accept that taxonomic classifications such as phylum and genus are largely arbitrary, but that species are a real category. They would not agree with Darwin's view that species are indistinguishable from variety.

Yet they have been unable to agree on a consensus definition, nor are they ever likely to do so.  This is a problem, obviously, if one is considering the process by which a species evolves. Consider, as a comparison, the common word "table." What is it?  As I write, I am sitting in front of a flat wooden surface supported by four legs. This is a desk, not a table – even if I should eat my lunch here.  Conversely, if I write at the kitchen table, it does not suddenly become a desk. Nor will the bench in my garage become a table or a desk just because I may occasionally eat or write on it.

A table is a real category, but the central fact about a table as a physical object is its intended usage, not its morphology. I think that the concept of species is much like the concept of table. We run into problems if we attempt to define species by a set of descriptive rules, and ignore function. Of course, we know what the function of a table is, but what could possibly be the function, or purpose, of a natural entity such as a species?

As a rule natural phenomena have no purpose. Rocks, rivers and mountains do not have a purpose – they just are.  Human activities, of course, often have a purpose. We build a table precisely to have a surface to eat from. The other place we see purpose is in biology.  We are comfortable with expressions such as: "The purpose of an eagle's talons is to catch prey."  In truth, however, the mutations that produced longer claws in the ancestors of the eagle did not happen on "purpose," but by chance.   Longer claws improved prey catching, and through natural selection they became more common.   Nevertheless, evolutionary biologists can safely use terms such as design, function or purpose to describe structures or processes that have practical use.

This is not to say that everything about a living organism has a purpose. Biologists often debate the reasons, if any, for the existence of a particular biological trait. The red pigment of the cardinal, for example, presumably has the purpose of attracting a mate. By comparison, the red color of tube worms that live at the bottom of the ocean has no purpose. Their color is an accident of chemistry, and has no function in the blackness of the deep sea.

The tendency to assume a purpose to biological traits has sometimes been overdone.   The Harvard University biologists Stephen J. Gould and Richard Lewontin famously criticized this tendency in their 1979 article "The Spandrels of San Marco and the Panglossian Paradigm: A Critique of the Adaptationist Programme."  Their critique may have been valid, but we should not abandon the search for adaptive purpose just because sometimes a purpose is claimed where none exists.

The issue of purpose has confounded speciation biologists. Some view speciation as primarily an accident of geography, with no adaptive function. Others believe that speciation is an adaptive process, and occurs so organisms can become more fit.

Ok, this is said badly. Evolutionary biologists can get upset by suggestions that organisms evolve for the purpose of enhancing their fitness. It's the other way around – enhanced fitness leads to the evolution of purposeful traits. Is speciation a purposeful trait, which enhances fitness?   If so, what is its purpose?  Perhaps the best answer comes from the influential evolutionary biologist Ernst Mayr, who had this to say about the function of speciation: "The basic biological purpose of the species is the protection of a harmonious gene pool."

Which leads us to ask, what is a harmonious gene pool, and how it is produced and protected?

## Genes in harmony

By "harmonious gene pool" Mayr meant the genes, and their variations, that endow a species with its unique adaptive traits. Consider the talons of the eagle again. Talons can be defined as long, curved claws with sharp ends. This definition, however, does not obviously eliminate the claws of a goldfinch, which may look like talons, but are used to hang onto stems. This is why dictionary definitions invariably include function when defining talons. Of course, an eagle is more than a bird with talons. It also has a hooked beak, sharp eyes, large body, a taste for fresh meat, and a host of other features that define an eagle. These reflect the "harmonious gene pool" of eagles. Because eagles only mate with other eagles, these harmonious combinations are preserved from generation to generation.

If we adopt Mayr's view of species, then organisms speciate in order to protect a set of harmonious genes . . .

OK, this is said badly again. More correctly, we should say that organisms that speciate protect a set of harmonious genes, and thereby have greater fitness than those that do not speciate. I doubt if Mayr ever said it in that way, but this is the logical conclusion of Mayr's view on the function of species.

This is important, so let me elaborate. Here is the logical sequence:

1. Preserving harmonious combinations of genes enhances fitness.

2. Speciation is a mechanism that helps preserve harmonious gene combinations.

3. Therefore, traits that promote speciation will enhance fitness.

I will have much more to say about those traits, such as sexual reproduction, that might promote speciation. For now, let's focus on defining what a species is.

I believe that a proper definition of species should include the traits that characterize a species as well as the biological function of a species. In doing so we may be limiting that which we call a species, but that is ok. We define talons by their function precisely to distinguish them from similar structures, such as the claws of a goldfinch.

To illustrate this point, consider how we classify animals of the genus Canis. Biologists, and ordinary people, classify dogs, wolves and coyotes as different species. We do so even though these animals can, and do, interbreed and yield normal, fertile offspring. Hybridization even occurs in nature. An analysis of the DNA of wolves and coyotes in the western US found that these animals occasionally mate. Interestingly, it appears that sex only occurs between male wolves and female coyotes. This makes sense. A male wolf who happens upon an ovulating female coyote is powerful enough to force himself upon her. A male coyote who happens upon a female wolf in oestrus is more likely to get killed than satisfied.

In spite of occasional hybridization, we accept that dogs, wolves and coyotes are distinct species. And they are. Anatomy and reproductive behavior can both be used as evidence for their uniqueness, but these primarily reflect a difference in their life styles. Wolves are social animals that hunt in packs to bring down large game. Coyotes generally hunt alone and kill smaller animals. Dogs, of course, have established a symbiotic relationship with humans, and would likely starve without us.

Each species within an ecosystem has its own special place in nature – its niche – and a unique harmonious combination of genes that enable it to survive in this niche. Each species must also have reproductive genes that help ensure the preservation of its ecological genes. This seems to be a problem with dogs, wolves and coyotes. Their ranges overlap, and they are both capable and willing to mate outside of their species. Such matings are rare, but they do happen. This is clear from studies of the red wolf, a "species" once common in the Southeastern United States. In 2011, an analysis of red wolf, coyote, gray wolf, and dog DNA revealed that the red wolf was about 75 percent coyote and only 25 percent gray wolf, so the red wolf is more coyote than wolf.

Is the red wolf a proper species? This is a question that does not have an answer. No definition of species could resolve a borderline case like the red wolf. We could, I suppose, test the degree of reproductive isolation by throwing grey wolves, red wolves, dogs, and coyotes together in a pen and observing their mating habits, but this is not a good idea. Blood is likely to flow, and it would tell us little about their behavior in the wild. The status of the red wolf as a species is still in dispute, and is likely to remain so no matter what definition of species is adopted.

Yet, we need a working definition if we are to proceed.

## A new species definition

Evolution is about winning. You do not have to be perfect, you only have to be better than the competition. In his prime, Tiger Woods had no peer at golf, but he would have had trouble making a living swinging a tennis racquet. Golf is Tiger's niche, and he is a winner because of traits that allow him to excel in that niche (Of course, it helped that he started at an early age, and had supportive parents). The niche is a species' arena of competition, and to the winner goes the prize of reproductive success. No blue ribbons or trophies, just offspring.

Species definitions rarely make any reference to niche. They emphasize anatomy and reproduction. There is a reason for this. Biologists can easily detect and analyze anatomical differences, so morphology is a natural place to begin if you need to distinguish one species from another. If anatomy is inconclusive, mating habits may provide a definitive result.

By comparison, no simple measurement can identify a species' niche. Yet it is the niche that has guided the evolution of those features that are common to the members of a species. Consider wolves and coyotes again. Are their anatomical differences the result of differences in their behavior, or *vice versa*? The answer is yes. Which is to say that anatomy and behavior evolved together, as a single adaptive complex. Certain features may emerge first, or be of particular importance, but fitness depends upon various traits working together in harmony.

Each species has its own set of specializations, which arise from its particular niche and which are maintained by its reproductive habits. Cheetahs run fast, elephants pick things up with their nose, giraffes nibble on tree leaves without leaving the ground, and so on for each species. Often, these specializations are not as obvious as the nose of an elephant, but they exist nonetheless. I cannot tell you, for example, what is special about the different species of sparrows that come to my backyard. The white-throated sparrow and the house sparrow are so alike that only careful observation can tell them apart. I am sure that their niche differs in some critical way, but the difference that matters is not readily apparent.   An ornithologist somewhere might know what is special about the life styles of these birds, but it is a mystery to me.  Yet birds can do something you and I struggle at – they can easily tell the difference between different species of sparrows – at least when it comes to identifying their own kind.

Species differ in both ecological genes and in their reproductive genes.  Ecological genes contribute to a species' fitness within its niche. These are the genes that determine how an organism acquires food and avoids being food for someone else. They also determine, to a large extent, what a species looks like and how it behaves. The talons of the hawk, the claws of the lion, and the teeth of the crocodile, for example, tell us that these organisms are predators, as does their taste for meat.

The other genes important in defining a species are "reproductive genes." These are the genes that help ensure that reproduction takes place between members of the same species. Reproductive genes control traits such as the tail feathers of a peacock, the song of the bullfrog and the dance of a fruit fly. They also control the reactions of potential mates to these signals. Included in this category are the genes that control the shape of an orchid's flower and the structure of the proteins that allow fusion of a starfish egg with a starfish sperm.

Reproductive genes evolved for the purpose of preserving ecological genes . . .

This, once again, is said badly. Like many of my students, I tend to simplify by assigning purpose to a biological trait. More correctly, reproductive genes that help conserve beneficial ecological genes increase fitness, leading to the spread of those genes. Please excuse me if I should error in this way again. I just can't help it! When I seem to assign a purpose to a biological phenomenon, like speciation, what I mean is that it enhances fitness.

This confusion about purpose in evolution is common. Let me quote Mayr again: "The biological meaning of species is thus quite apparent: The segregation of the total genetic variability of nature into discrete packages, so called species, which are separated from each other by reproductive barriers, prevents the production of too great a number of disharmonious incompatible gene combinations." Mayr is responsible for the most widely adopted definition of species, called the Biological Species Concept. In his words, "Species are groups of interbreeding natural populations that are reproductively isolated from other such groups." This definition has been adopted to apply to both sympatric and allopatric populations, but Mayr was perfectly clear that this was not appropriate.

This is important, so let me quote him again: "The isolating mechanism by which reproductive isolation is effected are properties of individuals. Geographic isolation therefore does not qualify as an isolating mechanism." Taxonomists, nevertheless, routinely classify geographically isolated populations as distinct species. This is ok, but these are not "Biological Species" as defined by Mayr, but "Geographical Species" (my terminology).

So here is my definition of species: *An ecological species is a population of organisms with reproductive genes that function to preserve a unique set of harmonious ecological genes.*

This definition is not intended to apply to everything that has been called a species, only to those that have speciated as a consequence of ecological adaptation. Bacteriologists, island ecologists and conservation biologists, among others, may need a different definition to include everything they want to treat as a species.

Obviously, one can criticize this definition, and that is ok. Nevertheless, it follows logically from Mayr's description of the function of species, and rejecting this definition requires, I believe, rejecting Mayr's insight. Of course, shared ecological genes mean that the population will display similarities in anatomy, physiology and behavior, and these can be used as criteria for the identification of a species. Because of their reproductive genes members of this population will tend to mate with partners who also have the shared ecological genes, which is why they are preserved. Here is Mayr again: "For a Darwinian to determine the significance of a biological process one always starts with the Darwinian why question. As far as the species is concerned, the answer clearly is protection of the gene pool through establishment of a reproductive community."

The proposed definition of species does have a serious limitation. Including function in the definition creates a problem for the practical taxonomist. How does one know that a population has undergone evolution of reproductive genes *because* it helped to preserve its ecological genes? This is a bit like looking at a sheet of plywood mounted on four legs and trying to decide if it is a table, a bench or a desk. The harsh reality is that systematic observation is the only way to resolve such questions, and even then a definitive answer may not be possible.

It is said, truthfully, that you can't be a little bit pregnant, or for that matter a little bit dead. It's all or nothing when it comes to these life transitions. But can one be a little bit speciated? There are, for example, no partial humans. You are either a member of the species *Homo sapiens*, or you are not. There is nothing in between. While this is true for humans, many other species are not so confined, and intermediate forms do exist. This is no surprise, as speciation is an evolutionary process that occurs by small steps over many years. Any species definition will always leave some populations in limbo, and provide a never-ending source of argument for taxonomists. The species definition that I have proposed, to be clear, will not change this situation.

No doubt, my definition of species will be attacked as too restrictive, unnecessary, impractical, and impossible to apply in nature. All valid criticisms, but ones that are shared by other species definitions. My interest is not in adding or subtracting species from a taxonomical list, but in understanding the source of Earth's biological diversity. Some species exist because speciation helped to preserve a harmonious set of adaptive genes. These are the species that interest me, because these are the ones that I believe will lead to an insight regarding the evolution of sex.

With a working definition in hand, we can now explore how one species becomes two.

# Chapter 5: Speciation

## Breaking up is hard to do

Organisms evolve in response to local conditions, but the resulting alterations in their ecological genes will not necessarily create a new species. Reproductive genes must also change. If some females evolve a preference for the color red, then the males with the same ecological genes as those females will need to become redder. How this might happen is at the center of the speciation problem. Reproductive choice appears to be a classic example of "stabilizing selection." In stabilizing selection the most favorable genotype is also the most common one, and deviations from the norm lead to a reduced chance of reproductive success. In a population of, say, red males and red-preferring females you do not want to be tending toward pink. Yet speciation requires precisely that kind of evolutionary change.

A critical step in speciation is the evolution of reproductive isolation. Mark Kirkpatrick and Virginie Ravigne, writing in 2002, estimated that about 100 mathematical models of the evolution of reproductive isolation had been published (I am sure more have been published since then). They simplified these models by sorting them into five categories (and several subcategories). The last element in their list is the presence of an initial condition, such as geographical isolation, that can lead to genetic divergences on its own. By this they mean allopatry.

Because allopatric speciation has dominated evolutionary thought for so long, it is worth looking into it in more detail.

## Allopatric speciation: A play in three acts

Allopatric speciation proceeds in three phases, as outlined below:

1: Find your own space. Speciation begins when a population becomes geographically isolated from others of its kind. Lakes and islands can lead to isolation, but isolation can also be created by geological features such as rivers and mountains. It can also be created by mutations that lead to changes in habitat preference. Isolation must be "just right" if allopatric speciation is to take place. The new habitat must be sufficiently isolated to prevent regular contact between the two populations, but not so isolated as to prevent immigration in the first place. This is why models of allopatric speciation invariably begin with a rare event, such as a storm blowing a flock of birds onto a newly formed island.

2: Drift and selection. Once a population establishes itself in a new habitat, it begins to evolve on its own, and evolution, by selection and genetic drift, occurs independently of the ancestral population. At some point, depending upon what definition is used, a new species may emerge. Or not. If the new ecosystem is like the old one, selection and genetic drift may lead nowhere in particular, resulting in the evolution of regional variations of the same species, called subspecies or races.

3: Getting back together. An isolated population may diverge from its ancestors enough to earn its own species name, but this cannot explain how hundreds of closely related species, such as the Hawaiian fruit flies, come to occupy a single habitat. Models of allopatric speciation allow for the eventual reuniting of two original populations. This event, called secondary contact, is thought to be an important pathway by which two closely related species begin living in sympatry. It is this step that is the source of much theory, and controversy.

Consider a situation in which a group of pioneers has become established in a new, isolated habitat. For convenience, let's call the ancestral population A and the recently isolated population B. Populations A and B undergo independent evolution for many generations, but eventually the two groups come back into contact. I assume that the populations still mate with each other, producing hybrids, which I will call C. If they do not hybridize, then populations A and B are already distinct species, and no further discussion is necessary. Theory and observation, however, indicate that hybridization is likely during the initial stages of secondary contact. At issue is the fate of the hybrids, and the consequences for populations A and B. If the hybrids are healthy, and as fit as either parent, the outcome is straightforward and boring – populations A and B interbreed and fuse, remaining a single species.

If A and B are sufficiently different, then the hybrids might have a problem. They will display a disharmonious mix of ecological traits, and may be at a selective disadvantage compared to the pure-breeding forms. Consider the example of two populations of cichlids living in isolated lakes. Let's say that the ancestral population was well adapted to eating snails, but the splinter population wound up in a lake rich in worms. The hybrids, which are not particularly good at eating either snails or worms, are likely to be less fit than either parent, and are unlikely to thrive.

If secondary contact is to lead to speciation, the hybrids must be less fit than either of the parental strains. This can occur, as mentioned, if the hybrids are ecological misfits. It may also occur because of hybrids are chromosomal misfits. In such cases the hybrids may die early, often as embryos, or be born sterile. The mule, the progeny of a donkey and a horse, is a well-known example of hybrid sterility.

To be clear, hybrid offspring may have either improved fitness or reduced fitness. Only the latter case is thought to lead to speciation, and is what I will focus on.


## Kissing your sister

The literature on hybrids is immense, but we can simplify things by considering two extremes: matings between close relatives, such as brothers and sisters, and matings between individuals who are not even members of the same species. At both extremes the fitness of offspring is reduced (if they are born at all). Maximum fitness occurs when individuals choose partners who are genetically neither too close nor too far apart. They are just right.

Sexual reproduction has evolved in response to both extremes. Many species, including our own, limit matings between partners who are genetically too close.  Human cultures, for example, have taboos against incest. This is why kissing your sister is not viewed as a normal romantic act. At the other extreme, every sexual organism avoids mating with a partner who is too distant genetically. This is accomplished through the reproductive isolating mechanisms that characterize each species.

But where do we place hypothetical matings between populations A and B, the new incipient species and the old ancestral one? These populations have been isolated for many generations, allowing them to diverge genetically, so they will not be closely related. If matings of A and B individuals yield viable, well-adapted offspring then no speciation event is possible, and we will say no more about this situation. Speciation is possible only if the hybrid offspring are less fit than their parents, either because they are poorly adapted or hybrids do not develop properly or are sterile.

Hybrid sterility often arises as a result of chromosomal incompatibility. A remarkable fact about chromosomes is that they are ridiculously unstable. Bits of chromosomal DNA (called transposable elements) can pick themselves up and move to other locations. Crossing-over, which normally just swaps genes, can be uneven, leaving one chromosome with more, or less, DNA. Chromosomes can break, turning one chromosome into two, or they can fuse, turning two chromosomes into one. Such chromosomal rearrangements can seriously disrupt the normal events of meiosis, and may result in eggs or sperm with either a deficit or surplus of some genes.  Either event can be catastrophic. Downs Syndrome, for example, which leads to severe mental retardation, occurs when a person is born with an extra copy of chromosome 23.  Clearly, too much of a good thing can be really bad.

The key to successful reproduction is chromosomal compatibility. If you have, for example, a pair of fused chromosomes, then you need to mate with a partner with the same fused chromosomes. If you don't, then the matching of chromosomes that is a key event of meiosis cannot occur properly, and your gametes are likely to have an excess or deficit of some genes. Even if the hybrids develop normally, they may grow up to be sterile. Hybridizing with a chromosomally incompatible partner is a risk, but what's a future parent to do? No genetic counselors are available to help choose a mate. Evolution is cruel, and when mistakes happen, such as the choice of the wrong partner, the outcome is predictable – the loss of your genes from the gene pool. Conversely a good choice enables your genes to proliferate, and anything that increases the odds of finding the "right" partner will improve fitness, and will be selected for.

An example of this phenomenon has been observed in Missouri tree frogs. Carl Gerhardt and his student Mitch Tucker studied two closely related species, the eastern grey tree frog, and Cope's grey tree frog. The two species look almost identical, but the eastern tree frog has twice as many chromosomes as the other species. According to Tucker and Gerhardt, the increase in chromosome number in the eastern tree frog has two important consequences – it increases cell size, and it slows down the tempo of the male's mating calls. The female's eastern tree frog prefers the slower tempo, which helps ensure that her mate will have compatible chromosome numbers.

We can look at this phenomenon in two ways: Chromosome doubling led to reproductive isolation (because of its impact on song tempo), which allowed a new species to emerge. Alternatively, speciation promoted chromosome doubling because it helped the new species preserve its new harmonious gene combinations. Like many topics in evolution, no easy answer exists to such a cause-and-effect question.

Finding a partner with good genes is one explanation for the extravagant displays of animals like the peacock. The quality of such displays may serve to advertise the bearer's health and overall fitness. This, however, is a different issue than finding a partner with ecologically compatible genes. Brightly colored feathers tell you nothing about the shape of the beak, or length of the talons, or (more critically) chromosome structure.

The quality of a potential mate's genes is generally unknowable. Nevertheless, a female might get lucky. She may choose a mate because of his color, or his song, or his aroma, and by chance also get a genetically compatible partner. This may occur rarely, but when it does it has important consequences. The offspring will be both ecologically fit, and also tend to have the parent's reproductive genes. They will be on their way to forming a new species.

Biologists have a term for this process – secondary reinforcement. Secondary reinforcement happens after secondary contact, but only under certain conditions. If they are met, a new species may evolve. Allopatric speciation thus consists of two distinct processes. The first occurs by geographical isolation alone, the second occurs by secondary reinforcement. This view has been expressed as follows by Conrad Hoskin, et al.

> Allopatric speciation results from geographic isolation between populations. In the absence of gene flow, reproductive isolation arises gradually and *incidentally* [my emphasis] as a result of mutation, genetic drift and the indirect effects of natural selection driving local adaptation. In contrast, speciation by reinforcement is driven directly by natural selection against maladaptive hybridization. This gives individuals that choose the traits of their own lineage greater fitness, potentially leading to rapid speciation between the lineages.

There are a couple of problems here. The first is the notion that reproductive isolation occurs "incidentally" as a result of mutation, genetic drift and the "indirect effects of natural selection." If this view is correct, most of the millions of species that occupy our planet are nothing more than random freaks of nature. I, for one, find this hard to accept. While evolution itself depends upon random events (mutation), it is made possible by the impact of natural selection. No selective force operates on reproductive choice in an isolated population, so allopatric models of speciation rely on pure randomness.

The second problem is more serious. Following secondary contact, hybrid offspring are assumed to be less fit. This is the selective force that is assumed to drive the evolution of new mate choice genes, and ultimately speciation. This assumption is bothersome. If hybrids are less fit, selection could also simply eliminate the less common variety. Indeed, this outcome is hard to avoid if secondary contact involves a small number of individuals.

Models of allopatric speciation invoke a series of unlikely events. First, a pioneer population establishes itself in an isolated habitat, which is neither too far nor too close to its homeland. Once established, it remains reproductively isolated for many generations. During that time, genetic drift causes the isolated population to become reproductively incompatible with their ancestors. Then, and only then, they reestablish contact with their ancestral population. If the most likely outcome of contact, which is extinction, can be avoided, they may become a new species. This process must occur hundreds of times to account for the diversity of species living in some lakes and islands.

The Achilles' heel of allopatric speciation theories is its reliance on genetic drift. In essence, this turns speciation into a chance event. This is not necessarily a negative, but it has a catch, call it Catch 23, that raises questions about the role of genetic drift in allopatric speciation.

## Catch 23

The catch in Joseph Heller's novel "Catch 22" is that you can't claim insanity to get out of the army, since wanting to get out of the army is in itself proof of your sanity. Models of allopatric speciation display a similar paradox. These models assume that neutral mutations cause an isolated population to diverge from its ancestors, but these same mutations are reproductively detrimental when the populations come back together. Some logical contortions are needed to explain how genetic changes go from neutral to harmful by the simple act of secondary contact.

Yet, plenty of evidence exists that separated populations can become reproductively incompatible, and in a relatively short time. What has not been proven, however, is that reproductive isolation is an *incidental* result of such isolation. Indeed, some studies suggest that geographic isolation alone did not lead to the evolution of reproductive isolation. A critical result was reported in 2010 by Roger Thorpe et al. who studied the reproductive behavior of anole lizards on islands in the Caribbean. They found no reproductive isolation had evolved between populations of lizards geographically isolated for up to eight million years. They found, however, that sympatric populations living in different habitats had become partially reproductively isolated. In their words: "This rejects the development of reproductive isolation in allopatric divergence, but supports the potential for ecological speciation, even though full speciation has not been achieved in this case."

Thus, a direct test of allopatric vs. sympatric (ecological) speciation has come down solidly on the side of the long-standing minority view. Even so, allopatric speciation remains the most widely accepted model, and will likely remain so for some time.

As an example of how ingrained allopatry is in biological thinking, consider a study of speciation from the rainforests of northeast Queensland, Australia. This is the home of the green-eyed tree-frog, *Litoria genimaculata*. It consists of two populations, a northern (N) and southern (S) one, reflecting long-term isolation of northern and southern rainforests during cooler, drier periods. The rainforests are thought to have reconnected about 6,500 years ago, bringing the two populations back into contact.

Like most frogs, the females of this species choose a mate based upon his song. For the southern female the correct choice is critical. If she mates with a northern male, her offspring do not survive past the tadpole stage. Oddly, the reverse mating, a northern female with a southern male, does not lead to the same problem. As expected, the frogs in the contact zone have begun to change. Males sing a different song there, and females have become more attracted to the new tune. As a result, a new species of frog has emerged in the contact zone. The authors of this study, a group from the University of Queensland, write: "... reinforcing natural selection has resulted in significant premating isolation of a population in the contact zone . . ." So far, so good, as this result is consistent with standard models, and the authors go on to conclude: "Thus we show the potential for reinforcement to drive rapid allopatric speciation."

On the surface, this work supports the standard models of allopatric speciation. Before making our final judgment, however, we need to consider an anomaly in the data. The frogs in the contact zone are not only reproductively isolated from the other lineage, but are also isolated "*incidentally* [my emphasis], from the closely related main range of its own lineage." In other words, the southern frogs now living in the contact zone, not only avoid mating with their northern kin, but also with their more closely related relatives down south. The authors never explain why they think the observed changes are incidental. Presumably they are adopting, uncritically, the common view that premating isolation arises initially as an incidental result of geographic isolation. In this case however, geography appears to have no role in a change of mate preference genes. Sometimes, even when data contradicts the allopatric view, it may be ignored or explained away.

Even if the initial stages of speciation are triggered by geographic separation, the final step occurs when the sister species are brought back into contact. But when two isolated populations come back into contact, they are *sympatric* in the contact zone. Geographic isolation may provide a kind of kick-start to speciation, but the final stages take place in sympatry.

This has to make one wonder how sympatric speciation ever got such a bad reputation. The short answer is that sympatric speciation has its own problems.

## What's in a name? Sympatric, Ecological and Adaptive Speciation

Sympatric speciation has been largely discounted for one reason. Biologists could not explain how it could happen at a reasonable rate. It was easy to see how a population in isolation could diverge from its ancestors. The problem was explaining how a single population could diverge when sexual reproduction creates genetic mixing between the new and emerging species. Consider the hypothetical fish described by the Kondrashovs. These fish were assumed to live in a lake with two kinds of prey, one large and one small. Selection should lead to two species of fish, one adapted to the larger prey, the other to the smaller prey. The fish, however, don't know this. They select mates by a trait, color, that is independent of size. So big fish mate with small ones, producing medium ones, even though a medium-sized fish is poorly equipped to consume either of the prey types.

For speciation to occur, gene flow between the diverging populations needs to be prevented. The Kondrashovs showed that this could happen purely by chance if the traits in question were controlled by several genes. Other evolutionary biologists attacked their conclusion, and argued that sympatric speciation would still be too slow to be important. I see no point in getting into the details of this argument, because I believe both sides have minimized an essential fact – sympatric speciation is a mechanism of ecological adaptation, and as such is subject to evolution by natural selection.

Throughout the biological word are many things that are highly unlikely (and impossible according to creationists). But evolution, working in its slow but sure way, works its magic. Every species alive today is the product of a series of unlikely events, and each speciation event selected those organisms *most capable of speciation*. The inevitable result of this process will be organisms with genetic architecture, whatever that may be, that promotes speciation.

Historically, the debate over speciation has raged over the role of geographic isolation, but biologists have begun to pay greater attention to the role of ecological adaptation in speciation, changing the focus from geography to biology. This change is essentially a revisiting of Mayr's biological species concept. Species, as he emphasized, function to preserve a "harmonious combination of genes."

Speciation of this sort is called either ecological speciation (defined as "the process by which barriers to gene flow evolve between populations as a result of ecologically-based divergent selection") or adaptive speciation (defined as the "processes in which the splitting is an adaptive response to disruptive selection caused by frequency-dependent biological interactions.") I prefer the term adaptive speciation, as it emphasizes that speciation is an evolutionary process. Strictly speaking, sympatric speciation is not the same as adaptive speciation. However, sympatric speciation is the outcome of an adaptive process, so the difference is not significant for the present discussion. Ulf Dieckmann and Michael Doebeli expressed the issue as follows in their 2004 book *Adaptive Speciation*:

> ... in adherence with our tenet that speciation research would benefit from concentrating on processes and mechanism, rather than on biogeographic patterns alone . . . , all the examples reviewed in this chapter must be recognized as representing instances of adaptive speciation. However, out of respect for the tradition of the field we retain, for the most part, the classic terminology of "sympatric" speciation.

In the remainder of this book I will use the terms sympatric speciation and adaptive speciation interchangeably. Keep in mind, however, that while sympatric speciation and allopatric speciation are distinct processes, adaptive speciation occurs independently of geography. Thus, we do not need two different mechanisms of speciation, one for sympatric populations and another for allopatric ones. Unless data is found that makes two mechanisms necessary, one will do.

It is worth emphasizing this point. Much ink and vitriol has been expended on the allopatry vs. sympatry debate. However, this dispute may be unnecessary, and a distraction from the real issues. Even when a species can be shown to have evolved under purely allopatric conditions, this does not tell us about the process of speciation. In principle, adaptive speciation can occur in an isolated population, and do so in a way that is fundamentally the same as that which occurs during sympatric speciation.

The key to adaptive speciation is the evolution of a harmonious set of genes, and their protection by assortive mating. Which is where sex comes into play. Sexual reproduction creates new combinations of genes, some of which may be ecologically more fit than the parental combinations. In this way sex can create new harmonious gene combinations. Think of sex as a genetic composer.

It is not enough to write the music, it must be protected. Like a mad composer who rips up his new symphony, sexual reproduction has the capacity to destroy the harmonious combination of genes it has created. The same process that brings genes together can also rip them apart. To understand sex, therefore, we need to examine how it enables organisms to protect their harmonious gene combinations.

## Finding the Right Mate

Mate choice is often discussed in the context of sexual selection, as exemplified by the grandiose feathers of the peacock. Evolutionary scientists, beginning with Darwin, have speculated upon the extreme traits that sexual selection can produce. As the saying goes, however, beauty is in the eye of the beholder. The grayish daubs on the breast of a male English sparrow seem rather tame compared to the peacock's tail, but a lady sparrow must find them quite attractive. Songs can also be beautiful, and the croak of the bullfrog is, no doubt, quite lovely to his potential mates.

Before I get accused of endowing dumb frogs with human emotions, let me elaborate. Science is about the observable universe. You may tell me that you find blond women (or men) attractive, but actions speak louder than words. A female bullfrog displays her "feelings" about a male's song by her attraction to it. Obviously, I cannot put myself inside the brain of a frog, any more than I can put myself inside your brain, and all we have to go on is behavior. So I am comfortable endowing frogs with feelings of "love" and "lust." These are, after all, just different levels of sexual attraction, which most biologists would grant to a frog.

Plants also engage in sex, but their sex lives differ from that of birds and mammals. Lacking legs and brains, they have evolved other means of finding a desirable mate. One could even argue that they have an advanced form of reproduction because they may induce others to do the hard work. A peahen uses her eyes and brain to evaluate the suitability of a potential mate. An orchid uses the eyes and brain of an insect to accomplish the same goal.

Many invertebrates, such as sea urchins and starfish, engage in sex at a distance. Unlike orchids, however, they have no help, and must rely on their motile sperm to find a willing egg.

A willing egg?

Indeed. An egg is coated with receptors that detect when a sperm has made contact. If the sperm has the right properties, the egg opens up and allows fertilization. If the egg and sperm are incompatible, no fertilization occurs. The fertilization proteins of eggs and sperm evolve rapidly, and exhibit tremendous diversity. In this sense, starfish resemble butterflies and orchids. We may not think of the signal molecules that coat the head of a sperm as being "beautiful" but in a potential mate (that is, an egg of the same species) they can elicit the loving response of fertilization.

Sexual organisms, all of them, go to great lengths to mate only with members of their own species. Hybridization between closely related species does happen, but it is risky and to be avoided if possible. For the benefit of your offspring, choosing the right mate is important.

Recessive genes, however, make it is easy to get deceived. A beautiful red-tinted male may nevertheless carry recessive genes for ugly blue. Dominant genes create hybrids that are like the dishonest lovers of fiction. They promise a rose garden, but give only weeds. So it is important to be as certain as possible about ones mate.

The recessive problem, because it creates gene flow between diverging populations, has been one of the main barriers preventing wider acceptance of sympatric speciation. The problem is alleviated, however, if the potential mate presents an honest picture of his genes. Such honesty is more likely when several genes contribute to a phenotype, in which case the displayed trait reflects an average of these genes. It was this insight behind the Kondrashovs' model of sympatric speciation. Traits controlled by several genes lead to a kind of "truth in advertising," in which a trait accurately reflects genetic makeup. This grants an organism a level of predictability about its offspring when it selects a mate. Of course, the plants and animals making choices about their mates are ignorant of this fact, but if they make a good choice their offspring will have a greater chance of success. Good choices mean more of their genes will end up in future generations, who in turn will make good choices. This is the logic of evolution.

The key to adaptive speciation is the preservation of harmonious gene combinations through mate choice. We tend to think of species in terms of their ecological specializations, as exemplified by the trunk of the elephant or the neck of the giraffe, but each species must also evolve reproductive specializations. Speciation is promoted when new adaptive genes and mate choice genes are transmitted together to one's offspring. Such coincident inheritance can occur in just three ways: Random correlation, chromosomal linkage, and so called "magic traits."

**Random correlation**: Evolution begins with the randomness of mutation. This bothers some people, who have trouble imagining how pure chance could lead to the evolution of biological complexity. Yet it does. The power of selection, acting on large populations over many generations, converts randomness into order. Chance can also link unrelated events.

You dream of winning the lottery and, the very next day, your number is drawn! This, of course, says nothing about the power of dreams to predict the future. With millions of people playing the lottery and dreaming of winning, somebody will have their dream come true. Similarly, random correlations between ecological and reproductive genes may be rare, but will eventually occur in a large population.

For example, as described by the Kondrashovs, a large female cichlid fish may be attracted to a red-tinted male, who by pure chance also large. Their offspring will tend to be large, with red-tinted males, and red-loving females. The offspring will, in turn, eventually mate, and they will be more likely than their parents to choose a red-tinted mate, who is likely to be larger than average. With each generation, the odds of large size and red color showing up together increases, and we are well on our way to the evolution of a large, red-colored species of cichlids.

Random correlation of ecological and reproductive genes is built into the Kondrashovs' model of sympatric speciation, but even without math it is clear that choosing a "good" mate can have a powerful impact on the success of offspring, even if full speciation is not attained. The Kondrashovs' model works because assortive mating can, in principle, increase fitness in as little as a single generation.

**Chromosomal Linkage**. Another way in which ecological and reproductive genes can be linked is . . . by linkage. A female that selects a mate with a particular trait, such as a red tint, is simultaneously selecting for genes that are chromosomal neighbors of the color gene. Suppose that genes that control body size and male pigmentation are nearby on the same chromosome. If so, a female who selects a red male is simultaneously selecting his size gene. Such linkage between ecological and reproductive genes is considered to be rare, and only a few examples are known.

However, these linkages may be more common that typically thought. Because linkage promotes adaptive speciation, they spread along with the new species that carry them. Processes that generate new linkages, such as chromosomal rearrangements, may also spread because of their ability to promote speciation.

This issue is crucial to what follows, so let me repeat it. Any trait that promotes adaptive speciation will itself have adaptive value, and will spread by natural selection. Thus a phenomenon that may seem unlikely, such as chromosomal linkage of ecological and reproductive genes, may actually be quite common. Indeed, chromosomal rearrangements during speciation may occur precisely because they help bring these genes together . . .

Sorry, I goofed again. What I meant to say was that a rearranged chromosome that links an ecological gene and a reproductive gene may improve reproductive success, and thereby become more common through natural selection. I discussed previously the "catch 23" of allopatric speciation, the argument that chromosomal changes are neutral in allopatry, but become harmful after secondary contact. Perhaps these chromosomal changes are not neutral, but are useful because they bring important ecological genes and reproductive genes into a linked state. In an allopatric population, speciation may take place more readily when a new linkage is established, and these will become important in preserving species integrity when the populations are reunited. Note that the same mechanisms that help drive allopatric speciation, can also operate during sympatric speciation. We do not need two different mechanisms of speciation.

**Magic traits**: On occasion, a single trait may affect both ecological fitness and mate choice. These have been called magic traits, and are considered to be rare. An example of a magic trait is found in Darwin's finches. Female finches select mates by their song. Like any trait, variation exists in their song preferences. Some prefer the high tones of a soprano, others the lower notes of a tenor. Males, of course, vary in their songs. Birds with long thin beaks produce higher tones than those with thicker beaks.

Birds, of course, do more than sing with their beaks, they also use them to eat. Thin beaks are better at probing for hidden insects, while thick beaks are better for crushing hard seeds. As a consequence, when a female selects a mate because of his song, she is also selecting him for his ability to eat insects or seeds. Selection will favor females who select a male with a beak like her own. Thus if she is thin-beaked and attracted to sopranos, she will have a better chance of producing successful insect-eating offspring. Ditto for the thick-beaked seed crushers who mate with tenors.

These choices, acting over generations, result in increasing divergence of the birds into two populations that differ in song style and eating habits. As the populations diverge, hybrids will become less fit (as they have a less useful intermediate-sized beak). The result will be two species where one existed previously.

Magic traits are a recognized pathway to adaptive speciation, but are considered too uncommon to be of great significance. However, in a 2011 article Maria R. Servedio, and coworkers, published an article with the title "Magic traits in speciation: 'magic' but not rare?" which questions this assumption. In their words:

> Speciation with gene flow is greatly facilitated when traits subject to divergent selection also contribute to non-random mating. Such traits have been called 'magic traits', which could be interpreted to imply that they are rare, special, or unrealistic. Here, we question this assumption by illustrating that magic traits can be produced by a variety of mechanisms, including ones in which reproductive isolation arises as an automatic by-product of adaptive divergence. . . . We conclude that magic traits are more frequent than previously perceived . . .

For many years sympatric speciation has been considered unlikely because the necessary ingredients seemed to be missing or rare. Yet, each of the three pathways described above may be more robust than usually thought. Random correlation of ecological and reproductive genes has been considered inadequate to drive speciation, but as shown by the Kondrashovs' it can drive speciation if the traits in question are controlled by several genes, which is probably true for most ecological traits. Linkage between ecological and reproductive genes has been considered too rare to be an important factor in speciation, but this view ignores the possibility that speciation itself has selected for such linkages. Finally, magic traits may also be more common than generally thought, and I will argue later that a particular magic trait may have had a critical role in the spread of sexual reproduction.

Each of the routes to sympatric speciation is enhanced in populations that display strong mate choice. Indeed sexual selection has long been recognized as a promoter of speciation, and it has been suggested that sexual selection alone may be enough to drive speciation. Sexual selection is typically viewed as a distinct mode of natural selection, one not driven by the usual mechanism of ecological adaptation. The peacock's feathers may enable it to attract a mate, but do nothing for his survival. Instead, they consume energy and make him more prone to predation. Consider, however, that peafowl would probably not exist as a species if it were not for the way in which they attract mates.

Thus, the main arguments against sympatric speciation are not the kiss of death to this process as many biologists seem to have believed. Furthermore, even if sympatric speciation is slow, it may still be an important evolutionary process, as it provides an important adaptive option, one not available to asexual organisms.

Keeping one's adaptive options open is critical for evolution. This fact has been demonstrated in a long term experiment using the bacterium *E. coli*. In a 2011 article, Richard Lenski, and colleagues showed that more adaptable bacteria prevailed over competitors that held a short-term advantage. They recorded evolutionary change over 52,000 generations of bacteria grown during nearly 25 years. He and his team compared the fitness of clones representing two genetically distinct lineages. One lineage eventually took over the population even though it had lower fitness than the other lineage that later went extinct. The winners likely prevailed because they had greater potential for adaptation. Lenski concluded, "In essence, the eventual loser lineage seems to have made a mutational move that gave it a short-term fitness advantage but closed off certain routes for later improvement . . . and the dead-end strategy allowed the eventual winners to catch up and eventually surpass the eventual losers."

So, yes, sometimes the tortoise really does beat the hare.

Presumably, you can see where this argument is going. Sex increases genetic variation and thereby speeds up adaptation to new ecological conditions, and this is the reason for its continued dominance. This is an old, and generally discarded viewpoint, but can, I believe, be salvaged by a closer look at sexual reproduction.

# Chapter 6: What is sex?

If we are to understand the evolution of sex, we should first agree on its definition. In 1998 President Clinton went on television and announced to the nation that he "did not have sex with that woman." This may have been the first time in history when people argued about what exactly "sex" consisted of. For most people sex refers to the act of copulation. If so, President Clinton's claim was technically correct, if disingenuous. He had engaged in a sexual act (oral sex) and that was sufficient, to many people, to make him a liar. In this view, sex is a particular kind of interaction between a male and a female, and a sex act occurs even if no possibility of pregnancy exists. Thus, the act of sex and reproduction can be viewed as distinct events.

Indeed, sex is not a mechanism of reproduction. It may be a precursor to reproduction, but strictly speaking, reproduction is what happens after sex. If you insist on treating sex as a means of reproduction, consider this: A key event of sex is fertilization, during which two cells, an egg and a sperm, combine to become one cell. Do the math. Two becoming one is hardly reproduction, it is the opposite of reproduction. I guess we could call sex a form of "deproduction." Nevertheless, it is ok use the term sexual reproduction. After all, sex is usually (not always) followed by reproduction.

Later, I will emphasize the importance of the first sexual organisms, which would have been single-celled. We know this because all multi-celled organisms alive today share the same basic sexual mechanisms, and presumably evolved from a common single-celled ancestor. For single-celled organisms sex is a rare exception to their normal mode of asexual reproduction by simple cleavage. Yeast, for example, switch to a sexual mode in response to stressful growth conditions. Sex, for them is a means of genetic mixing, not reproduction. Indeed, chromosomal mixing is the defining phenomenon of sexual reproduction. Not until the appearance of multicellular organisms did sexual mixing of chromosomes became a necessary precursor to reproduction in some species.

Sex can be conveniently divided into four events: To begin, the primary sex cells, called gametes, are made. In most species, the gametes consist of a large, immobile cell, the egg, and a smaller one, the sperm (or pollen), which is often mobile. Next, the sex cells must be brought into proximity, either by direct contact (the act of sex) or indirectly (with the aid of agents such as the wind or insects). The two sex cells then fuse (fertilization) to form a new cell that contains the chromosomes of the egg and the sperm. Finally, the newly formed cell grows and divides to produce a new, usually genetically distinct, organism.

The gametes are produced through a complex process called meiosis. Biology students typically are required to memorize its stages and terms such as prophase, metaphase, and anaphase, which in meiosis occur twice (in meiosis I and II). Adding to the confusion, lessons on meiosis typically include a discussion of crossing-over, which occurs during meiosis, but is a distinct process. Even worse, mitosis and meiosis are always described in the same chapter, even though they have different functions.

Meiosis is necessary for the simple fact that sexual organisms have two parents. You, for example, have 46 chromosomes, 23 each from your mother and father. To have children you need to reduce the number of chromosomes in your gametes back to 23. Otherwise, the number of chromosomes would double with each generation, which is obviously not sustainable. While you get 23 chromosomes from both parents, you get one copy, and only one copy, of every one of their genes (with the exception of the Y chromosome if you are female).

Accomplishing this goal is not trivial. Imagine, as a metaphor, that you are in charge of 46 children, consisting of 23 pairs of siblings. For a game, you want to create two teams of 23 children, consisting of one child from each family. You could walk through the playground, directing each child to go to one side or the other, but this would be time consuming. Instead, you ask the children to hold hands with their sibling. The 23 sibling pairs, still holding hands, are directed to line up in a row.

You then instruct them to drop hands, and walk in opposite directions. As a result, 23 children go to each side of the playground, consisting of precisely one of each sibling pair.

This is the basic idea behind meiosis. In the kindergarten metaphor we asked the children to hold hands as they lined up in the middle of the playground. Can chromosomes hold hands? In a way they can. Each chromosome has a special region (called a centromere) by which it attaches to its chromosomal sibling (called a homologous chromosome). In addition, crossing-over ties the DNA of homologous chromosomes into knots. The chromosomal pairs then move as a unit to the middle of the cell. Once there, the chromosomes let go of their partners, the knots of crossing-over are undone, and the homologous chromosomes move to opposite sides of the cell.

When the homologous pairs line up in the center of the cell, they do so without regards to their origin. One of each pair originated from the mother (the maternal chromosome), the other came from the father (the paternal chromosome). Returning to our metaphor, assume that the siblings consist of a boy and a girl, representing the paternal and maternal chromosomes. When they line up in the center of the playground, they do so randomly. Some girls are on the left, some are on the right. When they move to opposite sides, the two teams will contain a random assortment of boys and girls.

How many different teams can we form with our children? It turns out that 23 pairs of twins can create $2^{23}$ unique teams, which is 8,388,608. Likewise 23 pairs of chromosomes can be arranged in $2^{23}$ unique configurations of maternal and paternal pairs. Thus, 8,388,608 chromosomal configurations are possible in the human sperm or egg. During sex these are combined, creating 8,388,608 x 8,388,608 or 70,368,744,177,664 possible chromosomal arrangements. This is more than all of the people on Earth, even more than all of the people that have ever lived on Earth.

The different chromosomal configurations are of significance only if the maternal and paternal copies of each chromosome pair differ in some way. The homologous chromosomes have the same genes, but these genes are not necessarily identical. They can, and often do, differ. Which leads us to the issue of genetic variation and why sex exists?

## The Queen of Problems

Why sex exists, and is so common among living species, is a matter of considerable debate. In my evolution classes I explained the various theories about sex and finished with a disclaimer: "These are just ideas," I would tell my students. "Science still does not have a good idea why sex evolved and why it persists. This issue has been called the "Queen of Problems" in biology.

This, to me, is an astounding fact. Sexual reproduction is one of the most striking features of life on Earth. And we do not understand why it occurs. Thousands of articles have been written about the details of sexual reproduction, but comparatively little about why it exists and is so common. Evolutionary biologists can explain, for example, why a female praying mantis eats her mate following copulation (he is a good meal, and has served his purpose), but not why she needs to have sex in the first place.

The puzzle of sex is twofold: How it began, and why it persists. The origin of sex is a mystery – and will probably remain so. The first sexual organisms consisted of single cells that left no fossilized traces of their life style. So scientists can only speculate about the origin of sex. Rather than engage in unprovable speculation, biologists have focused on why sex is so common among today's species. Given the costs of sex, particularly the twofold reproductive cost that comes from having males, sexual organisms should be quickly displaced by asexuals. Some of the more prominent theories that have been proposed to explain the prevalence of sex are described in the next section. I will not critically evaluate these ideas, or the evidence for or against them. To do so would expand this book beyond reason. In any case, none of these theories seem persuasive. Furthermore, they are dedicated to explaining why males exist, and become useless when sex occurs without genders (yes, it happens).

## Theories of Sex

Explanations for sex come in many forms. Some are highly mathematical, others are purely conceptual. Most focus, in one form or another, on the genetic variety created by sexual reproduction. Some are widely known, and appear in the textbooks, others are obscure and hard to find even with a systematic search of the literature. I make no pretense that I know this literature well, or have found every relevant publication. Below, in my own terminology, is a summary of the more popular ideas, but many others could have been added to this collection.

**The tangled bank.** The name for this theory comes from the famous last paragraph of Darwin's *Origin of the Species*:

> It is interesting to contemplate an entangled bank, clothed with many plants of many kinds, with birds singing on the bushes, with various insects flitting about, and with worms crawling through the damp earth, and to reflect that these elaborately constructed forms, so different from each other, and dependent on each other in so complex a manner, have all been produced by laws acting around us . . .

The tangled bank hypothesis assumes that environments are complex and heterogeneous, and that room for success exists beyond that enjoyed by the parents. Offspring equipped with new genetic combinations may be able to exploit new, underutilized niches. This model has sexual organisms playing a kind of genetic lottery. They produce offspring with different genotypes in the hope that one or more of them will hit the fitness jackpot. Asexual organisms are more conservative, and prefer not to take this gamble. After all, most lottery players are losers.

The odds of winning a lottery are so small that you might as well throw your money into a hole in the ground. Yet somebody wins – and can win big. And that's critical. Most of us will never become a millionaire by doing our daily job, and a lottery provides a rare opportunity of attaining great wealth. In the tangled bank most organisms struggle to leave any offspring to carry on their genes. Asexual organisms are content with this meager but reliable return on their investment. Sexual organisms take the chance that they might get lucky and hit the fitness jackpot, leaving more offspring than they would otherwise.

**Reducing sibling rivalry.** A successful organism may produce many offspring, who will grow up to be in direct competition with one another for resources. They are unlikely to all be successful. Sexual organisms can reduce such sibling competition by producing offspring that utilize limited resources in different ways. This may lead to higher overall reproductive success, particularly in a heterogeneous environment, such as the "tangled bank."

**Disengaging Muller's ratchet**. Not all theories focus on the potential value of increasing genetic variation through sexual recombination. One popular theory is more about reducing variation. The variations that are reduced, however, are of a particular type – they are harmful. This might seem to be a minor issue, as evolution itself is a powerful tool for eliminating harmful mutations. An organism that carries a harmful mutation will be less likely to survive and reproduce, and the mutation should be quickly eliminated from the population.

This conclusion is valid, but only if others in a population lack harmful mutations. Evolution is a competition, and the winner does not need to be perfect, only better than its competitors. If everyone carries harmful mutations, then the one with the fewest mutations will have an advantage. But how could it come to pass that everyone carries harmful mutations? The explanation is called Muller's Ratchet (after the geneticist

Hermann Joseph Muller, whose work focused on the ability of radiation to cause mutations).

Suppose that each organism in a population is born with at least one new harmful mutation. Since everyone carries a mutation, the winners of the evolutionary competition will be those who carry the least harmful ones, but even they will carry at least one mutation. When these individuals reproduce, a new round of mutations occurs. After the second generation, each organism will have at least two harmful mutations. Again, the most fit of these reproduces, and have offspring with new mutations. This process continues with each new generation, and after a while even the most fit individuals will be loaded with harmful mutations.

As a metaphor, consider what happens in a football season. In the first game of a season one or more players on each team get injured (representing a harmful mutation). In the next game, additional injuries occur. As the season progresses ever more players are hampered by injuries. Each week the team with the fewest injuries is likely to win, but by the end of the season even the best team is beset by multiple injuries.

Now imagine that we are not required to keep the same players for each game. Instead, before each game we form new teams by randomly combining the existing ones. By chance some new teams will be injury free. Others will be burdened with extra injuries, but it is the team with the fewest injured players that will go on to win the championship. In a similar way, sexual recombination of genes enables parents to produce offspring who may lack harmful mutations.

Prior to sexual reproduction, half the parental chromosomes are discarded (by meiosis). This allows sex to function as a kind of mutation eraser. Sex can also eliminate favorable mutations, but this is not an issue since most mutations are thought to be harmful. Preventing the accumulation of harmful mutations provides a simple and powerful explanation for sex – if not for a fatal problem. For Muller's Ratchet to operate at least one new harmful mutation must occur in each generation. Actual estimates, however, suggest that for most organisms the rate of harmful mutations is much less than one per generation. Only relatively large, long-lived species, such as humans, come close to this rate.

While Muller's Ratchet cannot explain the preservation of sex today, it could have been important in the distant past. Modern organisms have elaborate mechanisms for preventing and repairing DNA damage, but early ones may have lacked these tools. Their higher mutation rate may have favored sex because it prevented the accumulation of harmful ones.

**The Red Queen.** The term for this theory comes from a scene in Alice in Wonderland in which Alice meets the Red Queen, who is constantly running but not getting anywhere. The queen explains: "Now, here, you see, it takes all the running you can do to keep in the same place." The Red Queen hypothesis views living organisms as engaged in a perpetual arms race between predators and their prey, and parasites and their hosts.

We tend to think of evolution as leading to adaptation to a particular environment, such as a desert or a lake. The environment of a species may change over time, but these changes are largely random and unpredictable. This is one reason the notion that sex enhances adaptation to changing environments has gotten little support. The ecosystem of a species, however, includes not just its physical environment, but also its (changeable) relationships with other organisms.

A good metaphor of the Red Queen can be found in American football (again). For each game the coaches vary their game plans, hoping to gain an advantage over the opponent. The offensive coach designs new plays to counteract the anticipated defensive strategy, and the defensive coach does the same, as he attempts to predict the plans of the offense. In theory, a team that did not change its plays regularly would become predictable, and likely become losers. The odd result of all this variation in play-calling is a game that basically stays the same from season to season.

Evolutionary biologists are divided over the validity of the Red Queen, but for now I leave you with this thought: If you were a football coach, would you stick with your best plays, the proven winners, or change plays in the hope of surprising your opponent? The best approach is not obvious. Nor is it clear that a successful organism can gain an advantage by randomly exchanging their winning genes with those of another organism.

**Bad neighbors.** Each gene is linked to others that are nearby on the same chromosome, and the fate of a gene depends upon its neighbors. As a metaphor, picture a chromosome as a street, and the houses on the street as the genes. Real estate values (fitness) in this neighborhood depend upon the quality of each home. Now imagine that a new owner (representing a favorable mutation) moves into one of the homes. He begins fixing up his property, potentially increasing the values of all the homes on the street. Unfortunately, his next door neighbor is a bum (a harmful mutation), and has let his property deteriorate. No matter how much money the good homeowner puts into his house, it will never attain its potential value. Indeed, the entire neighborhood will be economically depressed because of the one bad homeowner.

To get full value of his house the good homeowner would have to pick it up and move it to another neighborhood, leaving the bad homeowner behind. This is hard to do with homes, but it happens to genes as a natural consequence of sexual reproduction. By randomly shuffling chromosomes during sex, good genes can escape the influence of the bad ones. Of course, the opposite can happen, and sex can move a good gene into a bad neighborhood. This theory also requires that bad (genetic) neighborhoods are common, otherwise little is to be gained by moving.

Each of these theories has its supporters, and each has some degree of observational or experimental support. Nor is this close to a complete list of proposed explanations for the prevalence of sexual reproduction. However, none of the roughly 40 theories that have been published appear to provide enough benefits to overcome the high costs of sexual vs. asexual reproduction. A new approach may be in order.

Perhaps a solution can be found if we begin at the beginning ...

## The origin of sex – A case of cannibalism?

To simplify the discussion, let's ignore the occasional swapping of bits of DNA that takes place between some bacteria. While this may resemble sex, it is in many ways an entirely different phenomenon. Likewise, let's ignore viruses that can pick up genes from their hosts, and insert them into newly infected cells. Sex, in its modern form, is done only by *eukaryotic* cells, those with their DNA enclosed in a membrane-bound nucleus, and that is where we begin.

The first living organisms appeared about three and a half billion years ago. They did not engage in sex, and when it ultimately evolved it would have been unlike the form that exists today. A complex process like sex did not evolve in a single step. It must have evolved gradually, having elements that could lead to sex as we know it, but beginning as something quite different. While sex is a complex process, two phenomena are common to all forms of sexual reproduction – cellular fusion (fertilization) and chromosome reduction division (meiosis). But which came first? Logically, cellular fusion came first, as it doubles the number of chromosomes, while meiosis makes sense only if the number of chromosomes has previously doubled (Some theories begin with chromosome doubling, but they lead ultimately to the same outcome, and I will not discuss them here.)

Fusion of one cell with another is the essential step of fertilization, but may have begun not for sex, but for eating. Amoebas eat this way today in a process called phagocytosis. During phagocytosis an amoeba surrounds a prey organism, such as a bacterium, with projections of its cell membrane. These projections, called pseudopods, extend around the prey until they meet, at which point the membranes of pseudopods fuse. As a result, the prey organism becomes enclosed in a membrane bag (called a food vacuole) within the amoeba. Think of the process as a kind of lethal hug.

Amoeba-like organisms probably first appeared about a billion years ago. Among the other single-celled organisms common at that time were photosynthetic blue-green algae (the cyanobacteria). We know of them from their fossils, called

Stromatolites, which are the remains of huge reefs formed by these early life forms. About the same time that amoeba showed up on the scene the cyanobacteria began to decline in abundance and diversity, and by the start of the Cambrian (about 500 million years ago) had fallen to 20% of their peak. One explanation for the loss of Stromatolite builders is that they fell victims to grazing predators – the first amoeba.

Picture a shallow sea, about a billion years ago, and a Stromatolite reef under attack by recently evolved amoeba. In short order the surface of the reef is swarming with these single-celled predators. On occasion two amoebas meet, and things get interesting. The ability of an amoeba to perform phagocytosis depends upon the phenomenon of membrane fusion. Imagine a bacterium that has the misfortune to be attacked simultaneously by two amoebas. As one amoeba reaches around its prey, it eventually contacts the membrane of the other one. Instead of the normal fusion event – between projections of a single cell – we might get fusion of the membranes of two different cells. The result will be a single super-sized cell that contains two sets of chromosomes.

At issue is the fate of the fused cells. Because the combined cell has roughly doubled in size, it will proceed rapidly to division, creating daughter cells with twice the usual number of chromosomes. The fitness of these cells will depend upon the genetic similarity of their "parents." If the cells were too different, the offspring may suffer from genetic incompatibility. Each cell came with its own harmonious set of genes, and no good is likely to come from mixing these with a dramatically different set. If the cells are sufficiently similar, however, fusion may improve fitness. For example, a dominant gene from one cell could reduce the harm caused by a recessive mutation in the other cell. Some of these fusion cells would thereby have an evolutionary advantage, and increase in number.

Accidental cell fusion would have continued, creating cells with four or more sets of chromosomes. While having two copies of a gene can be beneficial, more than that becomes a burden, with little compensating advantage. In their blind pursuit of prey, primitive amoebas may have occasionally collided, fused and doubled their chromosome count. Then they did this again, and again, and again . . . Something like this happens today with certain kinds of slime molds, who begin as amoeboid cells, then fuse to form multi-nucleated slugs in preparation for reproduction. Selective pressure would have existed to eliminate the extra set of chromosomes, and thus something like meiosis would have evolved.

These early sexual amoebas would have enjoyed many of the benefits of sex as generally understood. For example, they would have been able to eliminate harmful mutations and adapt more rapidly to changing ecological conditions. So these sexually reproducing amoebas would have been evolutionary successes. Two amoebas could fuse, and gain a significant advantage in the process, but only if they were genetically similar. Strong selective forces would have worked against the fusion of dissimilar cells. Thus, one of the earliest events in the evolution of sex would have been the appearance of recognition mechanisms to restrict sexual union to cells of similar genotypes..

This is also an essential component of speciation.

So when sex first appeared something else, just as important, happened. Rapid, adaptive speciation became possible. The first sexual organisms lived in a world brimming with underutilized niches, waiting to be filled by organisms that could assemble and protect the right combination of harmonious genes. Indeed, these first sexual amoebas may have been uniquely positioned to diversify through adaptive speciation.

As I previously discussed, adaptive speciation is promoted by three possible means: random correlation of ecological and reproductive genes; chromosomal linkage of such genes; and "magic traits" that affect both reproduction and fitness. These early sexual amoebas would have been well positioned for speciation on all three counts. They would have existed in huge numbers, increasing the chance of random correlation of the necessary genes. They probably had few chromosomes, maybe just one, and some reproductive and ecological genes would have been physically linked. Crossing-over, which can break such linkages, may have been rare or nonexistent.

Finally, the cell membrane controls cellular fusions, but also controls traits such as the rate of movement, adhesion on different surfaces and the ability to identify potential prey. Thus, membrane structure would have functioned as a "magic trait" in speciation. Like beak shape in birds, which influences both the tone of their songs and their ability to consume different foods, the structure of the amoeba membrane would affect both ecological fitness and mating success.

Let me summarize the argument to this point. The first sexual organisms may have been a type of amoeba. Their ability to perform membrane fusion, which evolved for capturing prey, enabled them to fuse and combine chromosomes with another amoeba, the process that defines sex. Fusion between amoebae would be beneficial, but only if they were genetically similar. This would have led to the first sexually reproducing species. Speciation, in turn, enabled the amoeba to spread into new niches, and diversify more rapidly and robustly than asexual organisms.

The cannibalistic amoeba model for the origin of sex has been around for a while (I am not sure who first proposed it), but was criticized in a 2009 article by Root Gorelick and Jessica Carpinone of Carleton University. They wrote:

> There are three substantial problems with cannibalistic oral sex as the origin of sex. First, chromosomes with all their nitrogen and phosphorus are probably the most nutritious part of the cell . . . . Cannibalistic oral sex only appears to be adaptive if the organism eats the good stuff, especially because nitrogen and phosphorus are limiting nutrients in most ecosystems. Second, the imbibed cell is most likely to be a close relative, which may not be adaptive if accounting for kin selection . . . Third, cannibalistic oral sex does not explain how [meiosis] might have evolved.

This critique is, I believe, wrong on all three counts. And the errors are critical. An amoeba consumes prey by a series of coordinated membrane-based processes. An essential step is the fusion of pseudopods, which encloses the prey in a vesicle composed of a bit of the amoeba's external membrane. Fusion of the amoeba's membrane and prey's membrane does not normally occur, and would only happen if the two membranes were sufficiently alike.

When an amoeba meets a "close relative" they are likely to have the same membrane structure. In which case, the contact of the amoebas will result in the same fusion event that occurs when two pseudopods meet. This will result in fusion of the cells, but not the digestion of the internal components of either cell. Digestion only occurs when the prey organism gets engulfed in a phygocytotic vesicle (this is sometimes called a food vacuole). An amoeba will not consume a close relative, it will merge with it. Only if they were sufficiently different might both detect the other as potential prey.

This was not dinner then sex, this was dinner *or* sex. I imagine some ferocious microscopic battles took place as pairs of amoebas locked pseudopods in a fight to the death. But with the right "chemistry" the two amoebas would merge, and since they have roughly doubled in size quickly enter a cycle of cellular reproduction. The two sets of chromosomes would duplicate, and the resulting daughter cells would be diploid (having a double set of chromosomes). Although the "parental" cells were closely related (or they could not have fused) the individual genes would vary, creating new genetic combinations in the daughter cells. Just possibly, the result would be a new harmonious genetic combination – and a new species of amoeba.

In considering the early stages in the evolution of sex another important fact needs to be considered. In a 2005 article, Matthew Goddard et al reported the results of studies on the rates of adaptation of sexual and asexual populations of yeast. They wrote: "Our results indicate that sexual reproduction can provide a selective advantage during adaptation to new environments . . ." This is one of the few directs tests of the notion that sex can increase fitness in new environments, but there is more. They then noted: "In extrapolating from experiments with . . .single-celled organisms to multicellular organisms, at least two points need to be considered. First, yeast gametes are of equal size (isogamous) and hence the classical twofold cost of sex does not apply . . ." (The second point is not relevant to the present discussion).

When two single cells, of roughly the same size, have sex they do so by fusing their outer membranes. This produces a double sized cell, with two sets of chromosomes. After fusion, division takes place, and two new cells are produced. They differ genetically from the parental cells, but each is fully capable of continued growth and asexual reproduction. Thus, when two single-celled organisms mate neither one gives up significant reproductive capacity because of their sexual activity. Two cells enter into a sexual union, and shortly afterwards two genetically new cells are produced.

This is critical. A massive literature exists on the costs of sexual vs. asexual reproduction – the "cost of males" – but this only applies to multicellular organisms that have separate male and female genders. The first sexually reproducing cells were certainly single-celled, neither male nor female, and therefore not subject to the cost of males. The fusion of two amoebas was, to adopt a phrase, a kind of "same-sex marriage," but neither male-male nor female-female. The evolution of gender was still millions of years in the future, as was the cost of males. Furthermore, sex would have been a rare event. Most cells would have reproduced by simple fission, and sex would not have been an obligatory event prior to reproduction. Thus, single-celled sex was a relatively cost-free process. Many proposed benefits of sex, such as the elimination of harmful mutations, would have been available without having to pay a significant reproductive cost.

Life a billion years ago must have been precarious, as it is today. The primary organisms of the Stromalite reefs, the blue-green algae, would have been assaulted on all sides by predators. Evolution would have resulted in a variety of defensive tactics, which must have been very successful given the size of the reefs that formed. Then the amoeba arrived on the scene. The blue-green algae were not defenseless, but the amoeba had a new trick up their sleeve. One that the algae could not keep up with.

The amoeba could speciate.

Life on Earth is believed to have begun about 3.5 billion years ago. The oldest amoeba-like fossils come from rocks about 750 million years old. Complex multicellular animals became common roughly 200 million years later. Roughly speaking, it took about 2 billion years of single-celled life to evolve something into something like an amoeba, and only 0.2 billion years more for complex multicellular life forms to evolve. There is uncertainty in these numbers, but they suggest an explosion in biological complexity in a relatively short time, possibly driven by the appearance of sex and adaptive speciation.

This cannibalistic amoeba theory has sex beginning with cell fusion, which I think is most logical, but other pathways to sex have been proposed. All pathways to sex, however, lead to the evolution of mate recognition and speciation. In truth we will never know for sure how sex began, and the amoeba model is about as good as any – unproven and unprovable. My focus in this book, however, is the today, and why sex dominates among higher forms of life. Theories on the origin of sex do not resolve the modern issue – why does sex persist? It may have not been costly to those early single-celled organisms, but it is certainly costly today.

Not until the appearance of distinct male and female genders was sex a problem. These organisms were probably like the invertebrates of the seas today, which mate by releasing their eggs and sperm into the water. But why would females give up half their reproductive potential in this way?

## Male problems

The mathematics seems straightforward: Place a population of sexual organisms in direct competition with asexuals of identical ecological fitness, and the asexuals win every time. This occurs because the sexual females "waste" half of their reproductive potential producing males. If we eliminate the male-female aspect of sex, however, then sex becomes less costly. A population of hermaphrodites (having both male and female reproductive organs) is roughly the equal in reproductive capacity to an asexual population. In principle, a pair of hermaphrodites could meet, swap genes, and have offspring at a rate equal to that of asexuals. They would gain the benefits of sex, without its most significant costs. So why are hermaphrodites so rare?

For males sex is a win-win proposition. They commit minimal resources to the next generation and have the potential to fertilize the eggs of many different females. The cost of sex applies to females, not males. A male can even be thought of as a kind of parasite, one that uses a female to perpetuate his genes. She does nearly all the hard work, while often he contributes nothing more than the tiny cells that carry his genes.

Let's return to the primordial soup, and the early sexual organisms. Sex is merely, in this model, a side benefit of a predatory lifestyle. Because the cost of sex is small for single-celled organisms, the benefits need not be large for sex to become a common event. Sex, for these single cells, was like playing the lottery with other people's money. There was much to be gained, and at little cost. At this stage the issue is not sexual vs. asexual reproduction. These early "sexual" organisms maintained the capacity to divide and grow by the usual asexual process. The phenomena we associate with sex, cell fusion, doubling of chromosome number, followed by meiosis to reestablish chromosome number, occur only rarely.

As time passes something important occurred. To understand this we can look at the biology of a living organism, a rotifer *Brachionus calciflorus* that can reproduce both sexually and asexually. Recent studies carried out by Lutz Becks and Aneil Agrawal found that a heterogeneous environment favors the sexual mode. Habitats characterized by spotty availability of high quality food increase the frequency of sex, while homogeneous habitats favor asexual reproduction. This makes sense. In the heterogeneous habitat, selection should favor those individuals that can generate new combinations of genes, which may give their offspring an advantage. Sex also has the power to destroy favorable combinations, and it is through speciation that sexual organisms preserve new harmonious gene combinations. However, the rotifers studied by Becks and Agrawal show no evidence of speciation in response to a varied habitat. They enter the experiment as *B. calciflorus* and exit as *B. calciflorus*.

So how do they manage to preserve their hard-earned harmonious genes? The answer may tell us a lot about the early evolution of sex. Rotifers tend to stay in one place. They do not swim or crawl great distances. As a consequence, when one of them has sex, it tends to be with a nearby neighbor, who is also adapted to the local environment. So there is little selective pressure on them to alter their mating habits.

Like the rotifer *B. calciflorus* the early sexual organisms may have been able to switch between sexual and asexual modes of reproduction. But – and this is important – these species begin as sexual types. I know nothing about the evolutionary history of *B. calciflorus* but I can make one statement with confidence: The immediate ancestors of *B. calciflorus* were sexual. Today this species may have the capacity to abandon sex, but it owes its very existence as a species to sex.

Sexual organisms are like the plants that fill the sunny space created when a tree falls in the forest. Consider the raspberry, one of the first plants you might find in such a clearing. They grow rapidly, and commit large amounts of energy into producing sugary fruit. Does this make evolutionary sense? The answer, of course, is yes. Raspberries exist, and that fact alone tells us that their reproductive strategy is successful. The oaks and beeches will return to the forest clearing in time, and the raspberry will disappear, but with luck a few seeds from its fruit, consumed by passing birds, will find new clearings in which to sprout.

Early sexual organisms may have overcome their disadvantages compared to asexuals in the same way a raspberry overcomes its disadvantages compared to an oak tree. The oak tree eventually wins, but not before the raspberry gets its chance at reproductive glory. In a similar way, the existence of males, and their motile sperm and pollen, enabled those early sexual organisms to spread to new habitats. This metaphor is not perfect, of course. The forest is dominated by its slow growing trees, not the adventuresome raspberry. By comparison, the adventuresome sexual organisms dominate, not the steadfast asexuals.   Predicting the victor in a real-world competition is difficult, but we can be confident that early sexual organisms occupied a world of ecological and evolutionary chaos. The nimble sexual organisms adapted quickly, compared to their asexual neighbors, to newly emerging niches.

The notion that sexuals adapt more rapidly than asexuals is not new, of course, and is one of the first ideas that will occur to an undergraduate biology student. It is also one of the few ideas about sex supported by actual data. Yet, the process by which sex enables organisms to adapt more rapidly has confounded theorists.  The reason for this, I suggest, is that the theories failed to take into account the power of sex-based adaptive speciation.


## Shaw and the beautiful starlet

Not only is there no unifying theory of sex, most of the proposed ones, it seems to me, share a critical weakness. They fail to take into account the "Beautiful Starlet" problem. This is the term I use to describe a widely reported incident in which the writer George Bernard Shaw was introduced to a lovely young actress. Supposedly the actress gushed "Wouldn't it be great if we got married. Our children would have your brains and my beauty." To this Shaw is said to have replied, "Unfortunately, madam, our children would probably have your brains and my beauty."

The paradox of sex it that it both creates and destroys favorable gene combinations.  Shaw and the Starlet were rarities – individuals who had succeeded in very competitive fields. Maybe they would have children that are both beautiful and smart, but most likely they would be neither as smart as Shaw nor as beautiful as the Starlet. This is fine, and I am sure that they would have sired wonderful children, who would have been successful *in their own way*. Theories on the evolution of sex, in essence, offer up ways that this could happen.

Consider the issues that arise if science develops the ability to clone humans.  Let's say that you are given the choice of having yourself cloned, or of having children the old-fashioned way. Will you pick the predictable outcome of cloning, or the crap-shoot of reproducing sexually?

The "best" option is not obvious. Even if you happen to be a star in your own field, there is no guarantee that your children will grow up in the same economic conditions that made your success possible. Times change, as do the qualities needed for success. In nature success is never a sure thing, and for many it is an unattainable goal, they become one of the "living dead."

## The Living Dead

The most popular theories of sexual reproduction assume that the parents mate randomly. That is, each individual in the population is assigned the same chance of mating as every other individual. The equations derived through the various theories then tell us how much better their offspring might be when they use sex to reproduce. This assumption may be incorrect. Not all individuals have an equal chance of reproducing. Observations of natural populations indicate that most individuals leave no progeny to carry on their genetic legacy, while a few produce the great majority of survivors. Biologists have a term for these non-reproducers – The Living Dead. Although they may survive to a ripe old age, the Living Dead do not pass on their genes, and have no impact on evolution.

Consider the theory that eliminating deleterious mutation is one of the main benefits of sex. Individuals with harmful mutations, however, are exactly those who are least likely to reproduce. A similar problem infects most of the other theories. The Red Queen theory proposes that organisms that have attained reproductive age, and thus have avoided the ravages of parasites and predators, will produce more fit offspring by throwing away the very genes that enabled them to succeed. The tangled bank seems, at first glance, to favor genetic variety. Yet the survivors in the tangled bank are those who have succeeded within their small bit of the riverside. Why toss the genetic dice on the outside chance that your offspring will find success somewhere else.

Romantic literature is replete with stories of people (typically women) who move up in status by marrying above their station in life. Yet for every commoner that marries a prince, a prince must marry a commoner. Similarly, for every individual that finds a partner with better genes, another finds one with worse genes.

One thing is certain however, the prince and the commoner will have human children (Even if the handsome prince was once a frog!). The ecological and reproductive genes that are essential for the unique traits that characterize a species are relatively fixed. All humans, for example, are characterized by a set of genes that are uniquely ours. These are the same, or very similar, in all people. Which is why we all have upright posture, big brains and hands with opposable thumbs.

Of course people differ. We cannot all run equally fast, nor are we equally smart or dexterous. Compared to a chimpanzee, however, a typical person stands taller, thinks more logically, and makes more sophisticated tools. (On the other hand, a chimpanzee can climb a tree faster than any unaided human.)

Sex leads to different outcomes for genes that are fixed within a species, and those that vary. Children vary in many ways, but they all look human, and they in turn will have children who look human. Theoretical biologists have focused almost exclusively on the ability of sex to create genetic diversity, and have given little thought to the effort sexual organism put into limiting diversity.

The biologist Sara Otto has used a poker hand metaphor to explain this issue. Imagine two people who have been dealt strong, potentially winning hands. They are given the option of randomly exchanging cards with each other to create two new hands, or of keeping the ones they were dealt. They choose the latter option, of course. If they held losing hands, however, they would be happy to swap cards. In nature, two organisms that reach maturity and are ready to mate have, according to evolutionary logic, winning genetic hands. Why would they want to shuffle their cards?

Now let's modify Otto's poker metaphor, and give each player a high pair (aces in one hand, kings in the other). They are given the option of keeping the pairs, and exchanging the other three cards with an opponent. They would be happy to do this. How organisms can keep their positive genetic combinations, while exchanging their less valuable ones is, I believe, the key to understanding sex, and speciation. Sexual organisms do not mix all of their genes. They work hard to ensure that some gene combinations are preserved. This they do by selecting mates of their own species.

The issue here is balance – how can the benefits of constructing new genotypes outweigh the costs of disrupting successful ones? Before exploring this issue let's look at a similar one – the "benefits" of mutation.

## Sex vs. mutations

Some years ago the British dog trainer Barbara Woodhouse hosted a television show called No Bad Dogs. Her premise was that bad behavior by a dog was often as much the fault of the owner as of the dog. The problem, most often, was the interaction between the dog and the owner.  One does not have to spend much time in an evolution textbook before coming across the concept of "bad" genes, which reduce the fitness of individuals unfortunate enough to possess them.

To be clear, bad genes exist (as do bad dogs). A mutation that inactivates a critical protein is going to have detrimental effects on the organism. Mutations with such dramatic effects are rare, however. Most mutations have subtle effects, and whether they are "good" or "bad" depends upon their interaction with the rest of the genome.

The most common type of mutation alters a single location in a protein-coding gene.Because of the conservative nature of the genetic code many mutations do not alter the structure of the encoded protein. We will ignore these, and limit the discussion to mutations that change an amino acid in a protein into one of the other twenty amino acids.  General rules are impossible, and the impact of a mutation varies from gene to gene. Some genes are highly conserved, and almost any mutation is harmful. Most genes, however, vary from species to species, and even among members of a single species. This indicates that many mutations are relatively benign. This is to be expected. A typical protein may have hundreds of amino acids, but only a few are directly involved in protein function.

Changing one of the peripheral amino acids in a protein typically has only a small impact on its function, and an organism's fitness. Nevertheless, the common view is that organisms are well adapted to their environment, and any change, even a small one, is likely to be harmful. Environments, however, are never constant. Even if climate conditions are stable, ecosystems are in a state of flux. Parasites and their hosts, and predators and their prey, interact to create ever-changing conditions. As the environment changes, a mutation that was harmful may become beneficial. Rather than adopting the common view that most mutations are harmful, a more valid view may be that the value of most mutations depends upon ecological conditions. Furthermore, genes never act alone, and a particular mutant may be beneficial in one genetic background, and harmful in another.

Most theories on the evolution of sex start with the assumption that its benefits arise as a consequence of the new genotypes it creates. But sex is not the only such process that can do this – mutations also produce new genotypes. Discussing the benefits of sex is, thus, a bit like discussing the "benefits" of mutations. Teachers are likely to chastise students who believe that mutations take place "so organisms can adapt to changing environments." Praise, however, goes to students who provide the identical answer when asked about sexual reproduction. Yet, mutations and sexual recombination both produce new genotypes through an essentially random process, so why the different viewpoint?

Behind this discrepancy is the assumption that organisms have no choice when it comes to mutations, but that they do have a choice when it comes to sex. Mutations are viewed as an unavoidable consequence of a DNA-based genome, while sex is not universal and is a product of evolution. Yet this distinction may not be as profound as it seems. While organisms may be doomed to mutate, the rates at which they do so vary.

Theoretical and experimental studies have shown that mutation rates are subject to natural selection. Mark M. Tanaka and coworkers, for example, examined the fate of a (hypothetical) non-mutator organism. They found that in changing environments "there will be relatively long periods of time during which the *mutator subpopulation* can produce a *beneficial mutation* before the ancestral subpopulations are eliminated" (My emphasis). If we replace the terms "mutator subpopulation" with "sexual subpopulation" and the term "beneficial mutation" with "beneficial genotype", we may have a good beginning to an explanation for the evolution of sex.

To clarify, consider a hypothetical planet, call it Earth-2, on which lives organisms that never mutate, and which reproduce with perfect genetic fidelity. Most organisms on Earth-2, however, are mutators. Biologists on this planet would be faced with a dilemma: Mutations are generally harmful, and reduce the fitness of otherwise fit individuals, yet most organisms on the planet mutate. An astute biologist on Earth-2 would predict that the prevalence of mutators should reflect the rate at which mutations generate new more-fit varieties minus the rate at which these new forms revert to mutation-free status. As the calculations of Tanaka demonstrated, mutators can persist as long as beneficial mutations occur faster than new non-mutator strains are generated.

We might suppose that non-mutators on Earth-2 would gradually accumulate (because of their enhanced short-term fitness), and become the dominant form. However, even a well-adapted variety (mutator or non-mutator) can be driven to extinction by changing environmental conditions or plain bad luck. Only mutating organisms, however, would have the ability to reoccupy recently vacated niches. More time must pass before a mutation occurs that limits further mutations. Thus, the proportion of mutators and non-mutators or our hypothetical planet would reflect a dynamic relationship between rates of beneficial mutations, the appearance of non-mutators and the probability of extinctions.

Earth-2 biologists, and philosophers, would be faced with a dilemma: Should mutations be viewed as "beneficial?" On the one hand, they typically reduce fitness. On the other hand, they are essential for evolution, and create the variations needed for new adaptations. I can imagine students on Earth-2 being faced with the following exam question: "True or false: Mutations take place to promote adaptation in changing environments. Explain your answer in 25 words or less."

On Earth, the consensus answer is that mutations occur by chance and are generally harmful. While rare favorable mutations may promote evolution, this is not the reason mutations occur. Mutations are viewed as unavoidable errors that occur during DNA synthesis, and the (apparent) absence of non-mutators on Earth reflects an impossible evolutionary pathway.

Suppose we give our students this question: "True or false: Sex exists because it promotes adaptation to changing environments. Explain your answer in 25 words or less." Most biology teachers would accept "True" as an answer, or at least give partial credit.

Of course, important differences exist between mutations and sex. Mutations change genes, while sex merely recombines existing ones. What is important, however, is the impact of these changes on fitness. Mutations are generally harmful because they alter genes that are the products of millions of years of evolutionary fine tuning. The odds are slim that a random hit on a gene will improve it.

When two organisms have sex a similar issue arises. The mating pair have demonstrated their fitness by reaching adulthood. Yet, by reproducing sexually they ensure that their offspring will have neither parent's genotype. Theories of sex depend upon these newly assembled genotypes improving fitness, as opposed to reducing it. It is not easy to explain how two fit parents improve the fitness of their offspring by throwing away half their genes.

This, to me, is the ultimate paradox of sex. The various theories of sex tell us ways in which genetic recombination can be beneficial. They tend to ignore, however, that the same phenomenon that may enhance fitness can turn around and bite you.

All theories of sex begin with one basic fact. Organisms differ genetically. The causes of genetic variation are something of a mystery. Evolution would seem to lead to genetic uniformity, as only the most fit reproduce, leading presumably to a kind of genetic perfection. Consider, however, recent studies of Ram Maharjan and coworkers who subjected *E. coli* bacteria to conditions of near starvation (they were grown in low levels of glucose). After many generations, the original genetically uniform population had evolved into several distinct types. Some had altered genes for DNA regulation, others displayed changes in sugar metabolism, and some had evolved different membrane proteins.

Thus, the glucose-deprived bacteria had adapted to the new conditions by mutations in genes that served different roles in the uptake and metabolism of sugar. Genetic variation is apparently the norm, even under a single selective force. The genetic complexity shown by an evolving clone of asexual bacteria in the laboratory is presumably less than would be expected in a more complex organism living under natural conditions.

One can imagine the adaptive advantage an *E. coli* would have if it were capable of sexual recombination. In short order, the five mutational solutions to glucose restriction would be combined in a single super cell, which would quickly come to dominate the population. It is not that simple, however, as another important difference between sex and mutations needs to be considered. Mutations are essentially permanent, and a mutated gene rarely reverts to the original form. New genotypes produced by sex, by comparison, are short lived, as they are likely to be disrupted during the next round of reproduction. This is the Achilles heel of most explanations of the value of sex. Yes, sex can create new, useful genetic combinations, but it just as readily disrupts such combinations.

New adaptive genotypes can be reliably preserved, however, by appropriate restrictions on one's sexual partner. Such restrictions are essential steps in the speciation of sexual organisms. Indeed, as emphasized by Mayr, the basic characteristic of a (sexual) species are mechanisms for the protection of its "harmonious" gene pool. For any species, considerable genetic variation may exist, but the core traits that characterize the species are preserved from generation to generation.

It is this fact, I believe, that, may lead us to a fuller understanding of sex.

# Chapter 7: Sex and speciation: Solving Darwin's Dilemma

## Choosing a mate: The story of the stalk-eyed fly

Sexual reproduction involves choice. Always. In my backyard, several species of birds, and one species of mammal (the gray squirrel), take advantage of the wealth of food provided by my bird feeder. Three or four bird species may crowd the feeder simultaneously, jostling for space on the narrow platform. When it comes time to mate, however, they go their own way. Female cardinals go for the striking red headdresses of cardinal males, while the wrens are wooed by the love songs of their potential partners, and hummingbird ladies find the acrobatic aerial dance of the males to be a turn on.

At one time colorful cardinals, melodic wrens and acrobatic hummingbirds did not exist. Back then, the common ancestor of modern birds ruled the sky. The fossil record, and the hard work of taxonomists, gives us clues how these ancestral birds changed to produce the thousands of bird species alive today. Nothing is known, however, about how these primitive birds attracted a mate, or how these mating habits changed during the evolution of modern species.

When it comes to selecting a mate, sexual organisms have a primary, obligatory task and a secondary optional one. First they must identify a member of their own species. If more than one potential partner is available, they may also have the option of choosing which will be their mate. Selecting a mate of one's own species is critical, but little is known about this topic. By comparison, much is known about the biology of selecting among potential mates of your own kind. Let's call the first phenomenon "species recognition" and the second one "mate choice." One consequence of mate choice is sexual selection, which has been a popular subject of evolutionary research since Darwin. The distinction between species recognition and sexual selection has been studied in the stalk-eyed fly. The males of these insects have their eyes located on the ends of projections that may be longer than the rest of the body. Females prefer males with the longest eye stalks. The length of the stalks reflects overall health, so by opting for the best endowed males, the female is also increasing the chance of selecting a mate with good genes. This is a classic example of sexual selection.

In Hawaii two species of stalk-eyed flies have overlapping habitats, and will occasionally hybridize, but generally mate with their own kind. While females, of either species, prefer a male with the longest stalks, she begins her search of a mate by choosing one of her own species. Sexy eye-stalks are not sufficient. They must be attached to a male of the female's own species. Thus, the female has both a species recognition mechanism (which is not understood) and a mate preference based upon eye-stalks.

In most species, males are less discriminating than females. Sperm are cheap and plentiful, and often sperm are the only thing the male contributes to his offspring. On occasion, males will attempt to mate with partners who are not of their own species, or even their own genus. One of the more humorous examples of this was a rare Australian ground parrot that attempted to mate with a photographer's head (It may be relevant that he was wearing a green shirt, much like the color of the parrot.) Another, sadder, example was a Pacific sea otter that took a fancy for a seal pup, and drowned it while attempting to mate with it.

In spite of occasional mistakes, each species remains relatively constant over time. Their mate recognition genes help ensure that their harmonious ecological genes are preserved, and the occasional hybrid is either out of harmony, or fails to develop at all. So if all this effort is directed toward preserving these genes, how do new species ever form?

## The Red Queen gets going

The Red Queen is often compared to an arms race in which two warring countries strive to stay one step ahead of the other. As Mark Ridley has pointed out, however, the Red Queen differs from an arms race in that it is not progressive, but cyclic. Parasites and their hosts are constantly evolving in response to what the other side is doing. The weapons don't get better with each generation, they are merely recycled. Like the Red Queen's legs, the genes that control parasite and host interactions go up and down, but carry the species nowhere.

The Red Queen model is probably the most widely accepted explanation for the prevalence of sex. I suggest, however, that the Red Queen does not keep running in the same place. Occasionally she jumps ahead. Or at least her sister does. To understand how this may happen, I need to explore another controversial issue – group selection.

The expression "survival of the species" is sometimes used by students of biology and by equally uninformed narrators of nature movies. Biologists, however, generally agree that evolution acts on individuals, and their genes, not on a species as a whole. Rams, for example, do not engage in fights to enhance the fitness of their species, but to gain access to females.

Survival of the species is a version of "group selection," in which populations, rather than individuals, undergo natural selection. Group selection has largely been discarded as a significant component of evolution. The reason for this is straightforward. Evolution requires that two conditions are satisfied. The first is competition. Group selection could occur if one population is in direct competition for resources with a second one. This certainly happens, as when one species utilizes the same resources used by others within an ecosystem.

Competition between groups is not enough, however. The second requirement for evolution is that competition leads to the spread of those genes that lead to victory. This is a problem for group selection. If a group as a whole is successful then individuals within the group are also successful – regardless of their genes. Within the group, bad genes will get passed on along with good ones.

Yet, group selection, of a sort, does occur. Your body can be viewed is an example of group selection. The cells of the body work together as a group to ensure the survival of the thing we call a person. This works because the body's cells are genetically identical, and they live or die – and reproduce – as a unit. Even here, however, you can see the limit of group selection. Mutations may take place in a cell that cause it to grow, and evolve, on its own. It can become a cancer, and if unchecked will kill the very body on which it depends for nutrients.

Even when organisms of a group are not genetically identical, a kind of group selection can occur. A family group, for example, may cooperate to promote the survival of its members. In doing so, each individual helps protect the propagation of those genes it shares with its relatives. This is called kin selection. Consider my own situation. I have no children, and thus rank low on the fitness scale, but I am not quite one of the living dead. What keeps me from having zero fitness is that I have a brother and a sister who have had children, and their children have had children. Some of the genes that I share with my siblings have been perpetuated through my grand nieces and nephews.

A critical requirement of kin selection is that individuals recognize their close relatives. The extreme example of this is the care that parents devote to their offspring. In some species brothers and sisters, and even more distant relatives, may contribute to raising the next generation. A family can be viewed as a group that satisfies the conditions for group selection: They are in competition with other families, and members of the family share many of the same genes. What prevents family selection from being significant is that genes flow freely between different family groups. Indeed, many species have reproductive behaviors that serve to limit inbreeding.

Species resemble families. Like a family, members of a species share many of the same genes, in particular the "harmonious gene combinations" that distinguish them from other species.  Like a family, they recognize one another as relatives, which they do through the mate recognition genes that they have in common. Group selection on species was the topic of a 1989 article by Leonard Nunney of the University of California at Riverside. His paper, like a related one by Steven Stanley published 14 years earlier, propose that group selection on species can explain the prevalence of sexual reproduction. These two papers are almost never cited in other articles, reviews or books on sex or speciation. The reason for this neglect is not clear, but both Nunney and Stanley's work deserve more respect than they have received from the scientific community.

First, here is Stanley's 1975 view (Italics indicate my emphasis): "Species selection is analogous to natural selection but operates upon species . . . rather than upon individuals within species. It favors species that tend to survive for long periods or *speciate at higher rates*." And here is Nunney's 1989 expression of the same idea: "The net result of group selection is that sex is maintained because of its lower extinction rate (*or higher speciation rate*) and because asexual mutants only rarely arise."

One could argue whether sexuals have lower extinction rates than asexuals, but they certainly have a higher speciation rate. Indeed, as usually defined, sex is an essential element of speciation. To me, the arguments of Stanley and Nunney seem strong, and hard to dispute. So why have they been ignored for decades? They have not even been attacked (at least in the published literature that I am aware of), just ignored.

I can think of a couple of reasons. The first is that they link sex to speciation. At the time they were published, allopatry had a stranglehold on speciation theory. Speciation was believed by most biologists (as it as today) to require a period of geographical isolation, which has nothing to do with sex. Indeed, sex can be viewed as a barrier to speciation, as it leads to hybridization, and allopatry is invoked precisely because it eliminates the problem of genetic mixing between the old and the new species. If allopatric speciation dominates, higher speciation rates of sexual organisms, as suggested by Stanley and Nunney, could not contribute to the value of sex.

The second reason, I suspect, that Stanley and Nunney were ignored is that many evolutionary biologists have been the victims of a kind of logical blindness.  They concluded (implicitly) that because most species are sexual, then sexual organisms must have an advantage over asexual ones. But another logical connection can be made from this same fact – that sex enhances the formation of species – and this conclusion has been largely ignored. Even in the papers of Stanley and Nunney this connection is little more than an afterthought.

I believe also that evolutionary theorists were sidetracked by the few species that are sexual and appear to live alongside ecologically identical asexual cousins. They wondered why the asexuals don't take over, and this question has driven evolutionary theory for decades. However, the asexual types arose from sexual ancestors, and they exist only because sex enabled these ancestors to become successful species. Occasionally the asexual species do conquer their sexual relatives. The common dandelion, for example, reproduces asexually, but its flowers tell us that it had sexual ancestors.  So abandoning sex can be beneficial, and does happen on occasion. A few species are even able to switch between sexual and asexual reproduction. As a rule, they use asexual reproduction when times are good, but switch to sex when times get tough, or conditions change. Again, the asexual forms of these species exist only because their sexual ancestors existed, and had the ability to revert to an asexual mode of reproduction.

Stanley and Nunney suggest that the advantage of sexual types arises because they are less subject to extinction, but both allow for the possibility that sex wins because of it promotes speciation. These are actually overlapping ideas. Sexual species may *appear* to avoid extinction because they spin off a new species. Biologists assume that a population that survives an extinction event with little morphological change remains the same species. Quite possibly, however, it is a new species that survives. The old species is gone forever, but its progeny lives on. Here is how Nunney explained it:

. . . once some sexual lines become established, then group selection can act to maintain sex despite its short-term disadvantage. The short-term disadvantage is included in the model by assuming that, if asexual individuals arise by mutation within a previously completely sexual species, then the asexuals quickly displace their sexual conspecifics and the species is transformed to asexuality. The probability of this event is given by the transition rate . . . If the [transition rate] varies between lineages, then one of the effects of group selection is to favor groups (i.e., species) with the lowest [transition rate]. This occurs because lines that do convert to asexuality . . . are doomed to a high rate of extinction, and in the long term only those that do not convert to asexuality . . . survive.

I think Stanley and Nunney explain why sex is so common, although their ideas may need some modification. They see speciation as a phenomenon independent of sex. It is clearly not. Sex, as I have emphasized, both creates genetic diversity and preserves the harmonious gene combinations essential to survival.   These can act together to create new adaptive strategies – that is new species.

## Of Hawaiian crickets, elephant fish and butter hamlets

I propose that sex is common among Earth's species because in promotes adaptive speciation. It does this by creating new combinations of genes, some of which may enhance fitness, and preserving successful combinations through mate choice.   A prediction of this model is that the more robust mate choice is, the greater will be the *potential* rate of speciation. I emphasize the word potential, as actual speciation rates will depend upon many other factors, such as resource availability and preexisting competition.

Speciation rates are, like many aspects of evolution, difficult to measure in nature. The most rapid rate of speciation known is that of the African cichlids, which I discussed previously.  These fish fit the predictions, and it is becoming an accepted notion that speciation in these fish was primarily sympatric, and was driven by sexual selection.

Another interesting example comes from the island of Hawaii where Tamara Mendelson and Kerry Shaw measured the rate of speciation in a group of 38 species of forest-dwelling crickets, the *Laupala*. They estimated a speciation rate of 4.17 species per million years, a rate exceeded only by the African cichlids. There is an oddity about these species, however. They look alike, live in the same habitats, and eat the same food. They are distinguishable only by the songs of the males, and the song preferences of the females. The authors conclude that "divergence in sexual behavior may have caused this rapid speciation."

A couple of things are significant about this data. The time for full speciation is estimated at about 240,000 years. This hardly seems fast enough to compensate for the costs of sexual reproduction. And it is clearly not. Speciation is an extreme version of assortive mating, which must occur over many generations before full divergence occurs. Strictly speaking, the proposed benefits of sex arise not from speciation, but from assortive mating. Consider, for example, the Kondrashovs' model of sympatric speciation. Speciation begins with assortive mating, during which an ecological trait and a reproductive trait begin to evolve in concert. The benefits per generation are small, but lead gradually to full divergence. Evolutionary theory tells us one thing about the past that we can be confident about: Genes present in living organisms today exist because their ancestors who had early forms of these genes produced more offspring than those lacking them. Speciation of a Hawaiian cricket may take 240,000 years to reach completion, but it was already underway 239,000 years ago.

Even so, the advantages of assortive mating per generation are small, and clearly insufficient to overcome the costs of sex on its own. Nor did it need to be. What is important, as Stanley and Nunney emphasized, is that assortive mating leads to speciation, and it does this at a rate faster than an asexual variety can evolve.

Evolution happens in two possible ways, genetic drift and natural selection. Evolutionary theory accepts no other explanations. The Hawaiian cricket species are either an accident of genetic drift, or they represent the outcome of natural selection (or perhaps some combination of the two). Just about everybody likes to eat crickets, including birds, lizards, frogs and small mammals, and even some people. So although cricket food (primarily rotting vegetation) is abundant, reproductive success is not ensured. Mendelson and Shaw note that the crickets they studied look nearly identical, appear to consume the same foods, and that the different species often overlap. Males of different species will often be found singing next one another. They concluded that speciation in these crickets is acting on sexual traits, as opposed to ecological ones. That is, the only reason so many cricket species exist is that female crickets are picky.

This is jumping to conclusions. I am reminded of the debate between Darwin and Alfred Russell Wallace (who famously sent Darwin a letter describing his new theory of evolution). Wallace did not believe humans were solely the product of evolution. To him, the advanced civilization of England and simple life of the hunter-gathers of the jungle could not have been the product of the same selective forces. Darwin disagreed. Today we side with Darwin, and understand that the skills needed to survive in the wilds of Africa are more demanding than those needed to survive in the halls of Cambridge. Likewise, it is impossible, with current knowledge, to place ourselves in the world of a cricket, and to appreciate the subtle ecological variations that may affect its survival and reproductive success. Perhaps, as Mendelson and Shaw suggest, the different cricket species are merely different song stylists, but they may also be using these songs to help ensure the success of their offspring in a difficult environment.

A similar story of rapid speciation related to mating signals comes from the elephant fish of Africa. In their daily life, the elephant fish uses an electric field to locate food items in the murky waters in which they live, which it then collects with a tubular mouth that resembles an elephant's trunk. Like the cricket, the elephant fish "sings" to attract a mate. Their song, however, is in the form of electrical pulses. Like the Hawaiian crickets the elephant fish have used their ability to sing to diversify into several species, which to a casual observer look and act very much the same.

Different species of crickets or elephant fish may look similar, seem to eat the same foods, and appear equally susceptible to predation, but may, nevertheless, be ecologically distinct. Natural populations are extremely difficult to study. This reality has been demonstrated by studies of a colorful Caribbean reef fish known as the butter hamlet. The 13 identified color varieties of these fish appear to be incipient species, but it is not obvious if they are the result of sexual selection alone, or of an adaptive process.  Many hours of observations over 94,000 square meters of reef were needed to establish that differently colored fish had distinct behaviors. Oscar Puebla, who studied these fish for his PhD thesis, notes that "during a one-hour dive, you will observe on average color-based behavior for only six minutes . . . you have to look for the rare behaviors that may have a disproportionate ecological and evolutionary significance. You can't just take a snapshot of the species, you to spend long hours in the wild . . ."

The cricket of Hawaii, and the elephant fish of Africa, may also be the result of ecological adaptation. We may never know for sure, however, as most graduate students would rather do thesis research diving in the Caribbean than crawling through jungles or murky rivers looking for crickets or electric fish. What is clear from these studies, however, is that speciation within overlapping populations can be robust, and that sex, through mate choice, plays a significant role in speciation.

Mate selectivity is not, by itself, sufficient to create new *adaptive* species. Mate choice needs to be tied in some way to ecological genes, either through random correlation, chromosomal linkage or magic traits. Examples of these are hard to find, but a few stand out.

## The importance of pea aphids

Searching the literature for articles that link sex and speciation is difficult, as most deal directly with one issue of the other, rarely both. Nor is the connection always obvious. Consider the lowly pea aphid. David Hawthorne and Sara Via studied two "incipient" species of these insects. One that lives on alfalfa and the other on red clover. They analyzed the genetic architecture of traits that influenced ecological success and mate choice. These genes were found to be closely linked on an aphid chromosome. As discussed previously the efficiency of ecological speciation is enhanced if the genes for an ecological trait are inherited along with those of a reproductive trait. One way this can occur is if the genes are physically linked on a chromosome, as Hawthorne and Via found in pea aphids.

Hawthorne and Via conclude: "This type of genetic architecture may be common in taxa that have speciated under divergent natural selection." What struck me about this statement is that *all* species (exception for purely allopatric ones) are a result of divergent natural selection. Hawthorne and Via do not comment on how such genetic architecture (linkage of ecological and reproductive genes) is created, but chromosomal rearrangements during meiosis are one way in which this can happen. Hawthorne and Via also note that: "Natural selection is increasingly recognized as the main cause of speciation, yet the detailed ecological and genetic mechanisms of the process remain obscure." Maybe I can help. A mechanism well known to enhance speciation is sex. The changes in genetic architecture that enabled the pea aphid to speciate into clover and alfalfa specialists were possible because these insects reproduce sexually.

Chromosomal linkage between ecological and reproductive genes favors speciation, but this fact has been generally ignored by biologists. They assume that such genes are randomly arranged on the chromosomes, and unlikely to be linked. This ignores the possibility that selection has worked its magic, and helped preserve such linkages when they happen by chance. Organisms with linked reproductive and ecological genes are primed for adaptive speciation, and will more readily diverge to fill available niches. As a consequence, such linkages become more common over time. This is the logic of evolution by natural selection.

Closely related species often have different numbers of chromosomes, even when the number of genes is essentially the same. Why would this happen? Rearranging genes does not, as a rule, alter how these genes function. It does, however, alter linkage relationships. The traditional literature on speciation has, in my view, a very strange explanation for chromosomal changes during speciation.  The argument goes roughly as follows: During a period of geographic isolation, genetic drift leads to chromosomal rearrangements in one (or both) of the isolated populations. When the populations come back together, the chromosomal changes cause hybrid offspring to be sterile or otherwise less fit.  Selection against such hybrids favors the evolution of mate choice genes that prevent hybridization, which leads ultimately to speciation.

The problem with this explanation, as others have noted, is that selection would more likely lead to extinction on one of the populations, or selection against the problematic chromosome structure. Nor does this model explain why chromosomal changes occur in the isolated population.  As I discussed earlier, this is a kind of catch 22, requiring that chromosomal changes are harmless in isolation, but harmful in the hybrids.

Another possibility exists. Chromosomal rearrangements create new linkages, and these promote speciation. In the isolated population new harmonious gene combinations are evolving, as are chromosomal structures that help preserve these combinations. Anything that promotes adaptation through speciation will itself be selected for, including otherwise harmful chromosomal restructuring.

Speciation theory has been dominated by the notion that allopatric and sympatric speciation are fundamentally different processes. And they can be. A population in isolation will evolve by genetic drift and selection until any self-respecting taxonomist will want to give it a new species designation. It is also possible, however, that an isolated population will undergo sympatric speciation, splitting into two or more species as it adapts to the new environment.  However, only one of the species may survive by the time a biologist shows up. The new species will *appear* to have evolved by gradual change of the founding population, but may actually be the only survivor of multiple speciation events.

Is there a way to distinguish the two processes? Not easily, or definitively. I have been struck, however, by the differing results obtained when biologists look at the ability of individuals of isolated populations to successfully mate with their ancestors. It seems that some, such as anole lizards of the Caribbean, will mate with other isolated populations, while others, such as the green tree frogs of Australia, are less prone to do this. As a working hypothesis I suggest that the degree of cross breeding between isolated populations is a measure of allopatric vs. sympatric speciation in one of the populations. If they happily and successfully mate, then we are looking at pure allopatric speciation. If they do not mate, or produce sterile offspring, we may be looking at a case of sympatric speciation in one or both groups.

## Stuck on Sex

Many objections can be raised to the ideas presented here. Which is fine, as I have lots of my own. One of these stands out. I argued that sex exists because it promotes adaptive speciation by enabling organisms to both assemble and protect harmonious gene combinations. However, even if sex promotes adaptation through speciation, this may not be sufficient to preserve sexual reproduction in the face of its costs.  It has been estimated that a sexual population will be driven to extinction by an asexual competitor in about ten generations. This is much faster than even the most rapid estimates of speciation (about 30 generations). So it seems that speciation cannot be the driving force behind sex.

Asexual types typically arise from sexual versions of the same species with which it is in direct competition. An asexual may drive its sexual cousin to extinction, but new asexual species is likely to emerge from the existing one. Should the worst happen, and the asexual species becomes extinct its niche will most likely remain unused until a sexual variety evolves to fill it.

In attempting to understand why sex is so common, we can largely ignore many species we are familiar with. Birds and mammals always reproduce sexually. Even if they might gain by becoming asexual, they do not have a mutational pathway that can take them there. A handful of organisms can switch from a sexual to an asexual mode, and vice versa, and these have been the target of much research. These organisms, however, are interesting precisely because they are unusual, so we need to interpret results in their case with caution.

Go back in time to those (hypothetical) first sexual amoebas. They lived in a world ripe with open niches, ready for the taking by new species. Their membranes would have acted as a speciation-promoting "magic trait," able to influence both ecological success and mating habits. They presumably had small genomes, perhaps on a single chromosome, and many of the genes that controlled reproductive choice and ecological fitness would have been linked.

Put simply, these early sexual organisms were primed for speciation. The result of this is visible today in the huge variety of highly evolved, complex, single-celled organisms that occupy Earth today. Eukaryotic microbes are the most highly evolved, complex organisms on Earth. Yet, as sophisticated and complex as a paramecium is (and it is *very* sophisticated and complex), it is limited by its single-celled nature. Size matters, but the only way organisms can get really big is if they form multicellular cooperatives.

Becoming multicellular, however, creates problems of its own, including sexual reproduction. Entire organisms can't fuse, only single cells. But sex drove the evolutionary diversity that ultimately made plants and animals possible, and these multicellular forms, once they appeared, were not going to simply abandon sex. When those first sexual cells began to aggregate into multicellular organisms, they took their sexual behavior with them. Some of these early plants and animals may have abandoned sex, but if they did they are no longer with us. The survivors were the ones that stuck with sexual reproduction.

Consider the sponges, one of the most primitive forms of multicellular life, and how they reproduce. Here is a summary of sponge reproduction as described in a 1980 article by marine biologist Avril Ayling [For clarity, I deleted references from this quote]:

> Sponge species may have separate sexes or partially separate sexes with varying numbers of hermaphroditic individuals. Other species are hermaphroditic, producing sperm and oocytes simultaneously, sequentially in the form of protogyny and protandry, or alternating sexes Sequential changes in sex may occur during a single reproductive season or over consecutive years. Other characteristics of sexual reproduction that must be considered are: 1.) whether the sponges are viviparous with embryogenesis occurring within the adult, or oviparous with oocytes or zygotes shed into the sea, and 2.) whether sexual products are released synchronously or asynchronously in individuals or the entire population. In addition, a number of forms of asexual reproduction are found in sponges; for example, gemmule formation, budding, and fragmentation of the adult.

Well, that helps a lot!

Actually it does. First of all, sponges enjoy the benefits of sex, but at very little cost. They are fully capable of asexual reproduction, and its more rapid reproductive rate. Many species are hermaphrodites, and a single sponge will produce both eggs and sperm. As a result, they do not suffer the full "cost of males." A population of hermaphrodites can, in principle, reproduce at nearly the same rate as a population of asexuals.

From a reproductive standpoint, sponges rule! Which raises the question: Why are the seas not filled with sponges? Actually, the seas, and the lands, are filled with sponges. Ok, not sponges as we know them, but with the progeny of sponges. All of the animals of the sea and of the land, including you dear reader, evolved from *sexually reproducing* sponges.

Asexual reproduction may be a great way for sponges to create more sponges, but not for creating jellyfish, sharks and people. Only sexually reproducing sponges could do that. And they did. (Plants and fungi had their own sexual ancestors.)

Evolution is a historical process. What exists today, is here because of what happened long ago. Consider the QWERTY keyboard. It is not the best way to arrange the letters of the English alphabet for efficient typing, but we are stuck with it. To change it, millions of people would have to learn a new arrangement, and the potential gain is, apparently, not worth the trouble. Sex may not be the best way to reproduce, but each new species is a product of sexual reproduction, and turning sex off is not that easy.

At some point, most sexual organisms evolved the male-female dichotomy we are familiar with today. Nothing in sexual reproduction requires distinct genders, and the cost of sex largely arises from having males. So it is not the evolution of sex that is problematic, but the evolution of males. Most likely sexual genders arose because multicellular organisms had to deal with sex at a distance. Single-celled organisms typically live at high densities, and are able to travel long distances (relative to their size). Sponges spend most of their lives stuck on a rock, and could only continue to enjoy the benefits of sex by evolving motile gametes, the sperm cell.

The emergence of genders had its advantages for both sexes. Females gain because they can focus on producing eggs with a supply of stored energy to get their offspring through the early stages of development. Large, nutrient-rich eggs, however, are not efficient travelers. Males concentrate on producing the travelers. They make many small motile cells that can seek out the non-motile eggs. By relying on males for reproduction, females lose half their reproductive capacity. This puts them at risk of extinction from a more rapidly reproducing asexual variety. If one exists. However, the first species to fill a new niche would be sexual, and the asexual variety would only be formed by mutations in the sexual species. For sex to become dominant all that is needed is for the rate of speciation to be greater than the rate at which asexual reversions can occur. This is the essence of the model proposed by Nunny.

Today, most species are stuck in their sexual life style. A few can, under special ecological conditions, become asexual. These may give us some interesting insights regarding the role of sex, but it is worth remembering that most of these asexual types arose from a sexual ancestor. They too are the product of sex-enhanced speciation.

One possible exception to this rule is known, the bdelloid rotifers (the "b" is silent). Hundreds of "species" of these microscopic organisms have been identified. Yet they are purely asexual, and appear to have been so for more than 100 million years. The bdelloid rotifers present at least two problems for biologists: First, how have they survived for such a long time as asexuals?  Most asexual varieties are short-lived compared to their sexual relatives, and no other asexual varieties appear to have been around for as long as the bdelloid rotifers. Second, why are there so many species of bdelloid rotifers? Sex may not be required for speciation, but it helps. The issue was clearly expressed in 2007 by Diego Fontaneto et al.: "Asexuals are an important test case for theories of why species exist. If asexual clades displayed the same pattern of discrete variation as sexual clades, this would challenge the traditional view that sex is necessary for diversification into species." (A clade is a group of organisms that share a common ancestor.)

So what about the bdelloid rotifers? Can they diversify into new species without sex? Fontaneto et al. believe so: ". . .we show that a classic asexual clade, the bdelloid rotifers, has diversified into distinct evolutionary species. . . .The results show that sex is not a necessary condition for speciation."

This conclusion presents a problem for theories that link sex and speciation. The bdelloid rotifers, however, are a rarity, and we should not discard a theory based upon a single, poorly understood, exception. Perhaps there is something unusual about the bdelloid rotifers that has enabled them to survive and speciate for millions of years without sex. A possible solution to the bdelloid puzzle was published in 2008 by Eugene A.

Gladyshev et al. They studied a phenomenon called "horizontal gene transfer," which is the movement of genes from one organism to another by means other than direct descent (whether sexual or asexual). Typically horizontal gene transfer takes place when a parasite, such as a virus, picks up bits of DNA from a host, and carries that DNA to new victims.

The article by Gladyshev et al. called "Massive Horizontal Gene Transfer in Bdelloid Rotifers," reported that these organisms were unusual in that they carried genes from many other species. Most of the transferred genes in the bdelloids were of bacterial origin, although several were from fungi. This is very strange. There must be something in the bdelloid rotifer's life style that makes them subject to such gene transfers. Gladyshev et al. explained it this way: "It may be that [horizontal gene transfer] is facilitated by membrane disruption and DNA fragmentation and repair associated with the repeated desiccation and recovery experienced in typical bdelloid habitats. . ." The bdelloids are organisms that rapidly appear in a rain puddle, even in places where rain is a rarity. Apparently, as they await the next rain their cells can pick up DNA from the environment, and when they come back to life that DNA can become part of their own chromosomes.

If bdelloids can pick up DNA from entirely different organism in their environment, what about DNA from other bdelloids living nearby? Gladyshev et al. note: "Whether there may also be homologous replacement by DNA segments released from related individuals remains to be seen. If there is bdelloid rotifers may experience genetic exchange resembling that in sexual populations."

So the bdelloid rotifers, which seem to defy the usual rules of asexual organisms, may have simply found another way to have sex.

## Say it with flowers

The power of sex to create variety is nowhere more apparent than in the flowering plants. Many people are unaware that flowers are relative newcomers on the planet. The first flowers appeared about the time that dinosaurs were becoming extinct (I suspect that there is a relationship between these events, but that's another story). Before the evolution of flowers, sex among plants depended primarily upon the wind for the dispersal of pollen, as remains true for many plants today. Flowers, according to accepted wisdom, are beneficial because they enhanced fertilization. This makes little sense to me, and is another sign of the need for a different paradigm in thinking about sex.

It is not, I believe, that flowers enhance pollination, but that they enhance the *selectivity* of pollination. Plants that rely on insects are not necessarily more efficient than those that use the wind. The wind will always blow, eventually, but the desired insect may never visit your flower, and then proceed to another of the same species. Flowers are not necessarily about reproductive efficiency, they are also about mate choice. Yes, plants choose their partners. They do so by making use of matchmakers, which are the insects, and others, that carry their pollen.

Flowering plants may dominate for the same reason sex dominates. Flowers enhance the process of adaptive speciation. Consider the orchid, which comes in more than 20,000 species and grows in nearly all parts of the world. Orchids rely on at least two highly specialized symbiotic relationships: one with a species of mycorrhizal fungus, which provide essential nutrients to the orchid, and another with a species of pollinating insect. This extreme specialization should make orchids more vulnerable to environmental fluctuations than species with more generalized habitat requirements. Orchids are also the potential targets of a host of herbivores and diseases. Yet orchids thrive.

As in many issues in evolution we need to be careful in sorting out cause and effect. Are orchids successful because they have produced many species, or are there many species of orchids because they are so successful? I suggest that the first answer is more likely. Their highly evolved flowers enable them to be very selective pollinators. This makes it easier for them to generate new species, and to do so faster than their rate of extinction.

# Chapter 8: Sexy People

## The naked ape – ugh!

First a warning, the following paragraphs contain extreme speculation, and may be hazardous to your ability to believe anything I have written to this point. Proceed with caution!

In my defense, the literature on human evolution is replete with speculation. Indeed, when it comes to number of theories relative to actual data, the science of human evolution probably comes out on top. Most of these theories attempt to explain the reasons for our unique features. Our characteristic physical and mental attributes, such as upright posture, gripping hands, large brains and the ability to use language, have stimulated a great deal of thought. This is fun stuff, and is harmless if it is not taken too seriously. Speculations about the evolution of our violent nature, for example, should never be used to justify, or explain, wars and murders. We are not that simple.

We can be sure of only one thing: The emergence of early hominids involved more than a single morphological or physiological trait. Like all species, we are a consequence of multiple genetic changes. Otherwise, we would just be bipedal chimpanzees, or perhaps talking bonobos. Critical to our success as an emerging species was the ability to generate new, useful genetic combinations. Equally important was our ability to preserve such genetic combinations.

Here is a question you probably have never seen: Why do humans prefer to have sex with other humans? This may seem like a silly question – until you attempt to answer it. If you are a heterosexual male, what is it about a female human that appeals to you? If you are a heterosexual woman, what is it about a man that appeals to you? I am not asking if you are attracted to a well-shaped breast, or a square jaw, but why are you attracted to a *human* version of these anatomical traits.

Roughly six million years ago, humans began to split from the common ancestor we share with the chimpanzee. At this time, our ancestors were starting to walk upright, as evidenced by fossils of early hominid hip and foot bones. Other changes were taking place to our heads and our hands. We may have also been losing our body hair, although we cannot be certain as hair left no fossil traces. Other important changes also left no fossil record, because they involved not bones but behavior.

Consider the classic literary ape man, Tarzan. He was raised as an infant by apes, presumably chimpanzees. Yet when Jane appears in the jungle he is taken by her beauty, and ultimately has a child by her. As a fictional character, no rules restrict his behavior, but we can reasonably question this outcome. Shouldn't Tarzan have found a lover among the chimps?

The Tarzan experiment, as far as we know, has never been done, so we can only speculate. This is allowed, as long as the speculation is based upon evolutionary principles. Among birds, newly born chicks "imprint" upon the first moving thing they see, and treat it as a parent. Konrad Lorenz was famous for demonstrating imprinting, and would walk around with an entourage of ducklings in his wake. These birds may have treated Lorenz as a surrogate parent, but I do not believe these ducklings grew up with a sexual interest in humans. Birds attract mates through song and dance and fancy feathers, none of which Lorenz displayed.

Evolutionary biologists have long speculated about the anatomical changes that distinguish us from our primate relatives. Their focus of has been on adaptation, not sexual attraction. This is fine, but we only split from our primate ancestors when we stopped sharing genes with them. According to allopatric models of speciation, this occurred because we were geographically isolated from the other apes that were evolving into chimps. Otherwise, hybridization would have mixed human and chimp traits, and prevented the development of two distinct species.

Genetic evidence, however, suggests that hybridization between emerging human and chimpanzee species did take place. Human speciation may have been triggered by a period of allopatry, but the final stages seem to have occurred while the ancestors of humans and chimps were in occasional contact – and still found each other sexually appealing. At some point, however, the willingness of chimpanzees to have sex with humans faded, as did the desire of humans to do it with chimpanzees. This was good for both species, as there was little chance of a chimp-human hybrid being successful. Today the two species have different numbers of chromosomes (chimps have 24 pairs to our 23) and hybrids would most likely be non-viable.

Speciation requires the evolution of mate recognition genes, and we should look at these genes to fully understand how humans emerged. Suppose, for purposes of argument, that evolution of a more upright posture was the single most important anatomical change taking place. This assumption is consistent with fossil evidence that bipedalism was an early step in human evolution. Indeed, evidence of bipedalism is what defines a hominid in the fossil record. Plenty of ink has been spilled speculating how bipedalism may have increased ecological fitness. Here is a short list of some of these ideas, none of which are supported by much evidence.

Bipedalism made it possible to see longer distances on open savannas, allowing easier detection of prey, and enemies.

Bipedalism made it possible to stand in shallow water while collecting shellfish.

Bipedalism freed the hands for tool use.

Bipedalism increased the efficiency of long distance pursuit of prey.

Bipedalism allowed food to be carried to a secure location for consumption and sharing.

Bipedalism made it possible for mothers to carry infants, which made it possible for them to be born at an early developmental stage, before they could cling to their mothers.

You may want to add to this list, but we will never know for sure why our ancestors began to walk around on two legs. Keep in mind, however, that a chimpanzee that walks on two legs is not more fit. It is just a poorly designed chimpanzee. Humans are more than a bipedal chimpanzee.

Consider a hypothetical population of hominids in which some are nearly bipedal and some walk with a slouch. For various reasons (see the list above) fitness is higher for more upright hominids. We expect evolution to gradually increase the frequency of bipedalism. However, there is a problem. Nearby lives a group of chimpanzee-like primates. Bipedalism for them is worse than useless, as it makes them less agile in the trees. If the two groups interbreed, neither will attain the maximum fitness appropriate to their niche.

This is no problem for allopatric theories, which is why they have remained so popular. Comparison of chimpanzee and human DNA, however, suggests that hybridization did occur as these species began to diverge. This makes sense to me. After all, these animals were not living on islands. One group may have spent more time in the savannah, and the other in the forest, but no impassible ocean separated them.

Whether speciation of humans was triggered by a bout of allopatry, or not, eventually hybridization ended through the evolution of mate choice genes, *in both species*, which made humans and chimpanzees repellent to each other.

Ok, "repellent" might be a bit strong. But mate choice may be driven by more than attraction. Indeed, people are not only attracted to certain physical traits, they are repelled by others. The same goal, assortive mating, can be accomplished by either response. A slouching, hairy body may be enough to make us look elsewhere for a desirable mate.

To make the concept concrete, imagine an early hominid, call him Moog, walking (he is a biped) through the jungle. Also present is an early chimpanzee ancestor, call her Cheeta, strolling along on her feet and knuckles. Cheeta is ovulating, and anxious to find a mate. Moog is a male, and always interested in sex. They meet in a clearing in the jungle, and pause to evaluate the situation. Although they are only a few feet apart, Moog, being upright, is unable to detect the delicate aroma that indicates a female in oestrus. He knows from bitter experience that attempting to copulate with an unwilling female can result in bites, scratches and frustration. In any case, she is not very attractive. Her hairy body and bad posture actually makes her somewhat repellent. Cheeta is in the mood for sex, but the male is showing no sign of interest. She turns her rear end toward him and wiggles it, but he does not respond. This is probably just as well she decides. He is almost hairless, probably a sign of parasites or skin disease, and his arms do not even reach the ground. She gives a grunt and shuffles back into the forest.

Moog walks back into the open savannah, and there he comes across Lucy. This is more like it! Lucy has excellent posture, and smooth almost hairless skin that appears free of blood-sucking parasites.  Lucy also admires Moog. She finds his upright stature a turn-on, and instinctively knows that he will be a good father for her children. Their matings will give birth to offspring prone to bipedalism and a preference for a bipedal mating partner.  Because bipedalism increases fitness (see possible reasons above) these offspring will more likely survive, and have offspring in turn.

The rest, as the saying goes, is history.

This, of course, is sheer speculation. The only "evidence" I have found related to this speculation is the story of Oliver the chimpanzee. Oliver was acquired as a young animal in 1960 by trainers Frank and Janet Berger of New Jersey.  Physical and behavioral traits of Oliver led the Bergers to believe that he may have been a human-chimp hybrid.  Oliver possessed a flatter face than most chimpanzees, and was habitually bipedal, never walking on his knuckles like a chimpanzee. There were reports that other chimpanzees tended to avoid him, and Janet Berger reported that Oliver became sexually attracted to her when he reached the age of 16.  After he tried to mount her several times, he was deemed a potential threat to Janet, and was sold. He spent the remainder of his life as a scientific and public oddity. Ultimately, DNA analysis indicated that he was not a hybrid, but pure chimpanzee. (I suspect that he may have been a bonobo-chimpanzee hybrid, but as far as I know this was never explored. Bonobos, also known as pigmy chimpanzees, are more sexually active than chimps and more prone to walk upright.)

Oliver may have preferred human females over chimpanzee females, but why is unknown. Nor is it clear that female chimpanzees found him sexually unappealing.  It would have been interesting to know if bipedalism was involved. During the early stages of human evolution bipedalism could have served as a "magic trait," with an impact upon both ecological and reproductive success. I don't want to be accused of over interpretation, but it is interesting that the most famous ugly person in literature is the Hunchback of Notre Dame. I also find it interesting that when chimpanzees put on dominance displays they tend to run on two legs.

Of course, we are more than upright chimpanzees. I once asked a group of women what they found most sexually appealing in a man. I expected them to cite anatomical traits that make a man "handsome," such as big muscles or a square jaw. Certainly this is kind of the response you would get if you asked a group of men what appealed to them in a woman. Surprisingly, they agreed on the single most important trait that made a man sexually attractive – his intelligence.

The traditional view is that our big brains resulted from ecological adaptation, but sex may have also played a role. Millions of years ago our female ancestors may have been attracted to males with signs of smarts. Perhaps a large head, relative to body size, could indicate intelligence. Females choose these primitive Einsteins as mates, and here we are today.

Rapid adaptive speciation is promoted when one gender, usually females, are selective about their mates. This selectivity makes it easier to combine key reproductive and ecological traits in the same genome. Did early human females have their pick of possible mates? Could she reject a dominant male for one, say, who displayed exceptional intelligence? And how would she know that he was a hominid Einstein? Matt Ridley, in his book *The Red Queen*, reviews some of the extensive scientific literature on what men find attractive in women, and *vice versa*. This is fascinating stuff, but may have little to do with either human evolution or sexual selection. The key to sexual selection is differential reproductive success. Ridley writes that men tend to choose beautiful women, while women choose dominant men. As he explains it: "If for most of human history beautiful women and dominant men had more children than their rivals – which they surely did because the dominant men chose beautiful wives, and together they lived off the toil of their rivals – then in generations women became a little bit more beautiful and men a little bit more dominant."

What is striking about this is the assumption that "for most of human history" men had a choice, and that they would be that particular. Meanwhile, the women also had their choice to make. All of this choosing supposedly leading to enhancement of male dominance and female beauty. Compared to our nearest primate relatives, human males are closer to females in size, suggesting that male dominance played a minor role in our evolution. The size difference between males and females may reflect different ecological roles rather than dominance relationships. Males did the hunting, and females the gathering. In any case, modern views of beauty and dominance do not explain how humans diverged as a new species.

Sexual selection probably played a role in shaping human evolution. Yet it is difficult to find clear examples. For example, the notion, as described by Ridley, that Scandinavians tend to be blond because of a male preference for blonds is hard to accept. To men, all women are sexually desirable, though some are more so than others. Brunettes, as far as I know, do not have lower reproductive success than blonds, even in Sweden. To argue that blond hair is a sexually selected trait requires two things – that some people prefer blonds, and that they will have more offspring if they go for a blond. The first statement may be true, but the second is much harder to prove.

Men may be attracted to blonds, but they do not throw the brunettes out of bed.  Hundreds of articles, popular and scientific, have been published on what men and women find attractive in the opposite sex. The consensus view is that people are attracted to a potential mate who demonstrates "good genes." One such measure of goodness is facial symmetry. Both men and women rate symmetrical faces as more attractive than those with an unbalance in their features. This has been well documented, but this does not tell us why men and women prefer another human over a symmetrically perfect chimpanzee.

People display several odd traits that *may* be related to our sexual behavior. The distribution of our body hair is certainly weird. Pubic hair is a sign of sexual maturity, but it is unknown if this is significant, or just an oddity of development. It is certainly not a good idea to have sex when you are not yet ready.  A patch of hair, strategically positioned, may have evolved as a visual signal of one's sexual maturity. Or maybe not.

While standards of beauty are interesting to speculate about, they may have had little impact on evolution. Males are not that particular and will have sex with even unattractive women (I know this for a fact). For most species, females tend to be more selective than males, but we do not know if this were true for our early ancestors. Some men may prefer big-breasted blond women, and some women square-jawed, blue-eyed men, but the species *Homo sapiens* is not defined by these traits.

As humans continued to evolve, we went through many changes. Heads got bigger, teeth smaller and feet flatter. According to some scientists, we also split into more than one species, all but one of which has become extinct. Not long ago (in evolutionary terms) we shared space with the Neanderthals. Genetic data, based upon a few usable samples of DNA, has been inconclusive, but it is generally believed that modern humans (Cro-Magnon man) and Neanderthals were different species. If they did interbreed successfully, it was rare. A recent analysis, led by Andrea Manica, from the University of Cambridge indicates that modern humans and Neanderthals rarely interbred. In her words: ". . . if any hybridisation happened – it's difficult to conclusively prove it never happened – then it would have been minimal and much less than what people are claiming now." This conclusion has been disputed, but in my view makes sense. Indeed, we can deduce this not only from advanced DNA analysis, but from the simple fact that their fossils are so easily distinguished ours.

Why was interbreeding rare? Did Neanderthals find Cro-Magnons ugly, and vice versa? If so why? We are once again reduced to speculation. I am struck by images that scientists have constructed of what Neanderthals looked like (These are easily available on the internet). Personally I did not find these imagined Neanderthals particularly unattractive (you may have a different reaction). More interesting was the effort of Dr. Robert McCarthy, an anthropologist at Florida Atlantic University, to recreate the voice of Neanderthals. They may, or may not, have had fully formed languages, but fossil evidence suggests that they sounded very different from us. Their voices appear to have been higher pitched, and they would have had trouble producing true vowel sounds.

They may have had language, but it would have been alien to our ears. Maybe even repulsive. Indeed, recent research has found that women prefer men with deep voices. Furthermore, males with deep voices tend to have sex with more women than do men with high pitch voices, and to have greater reproductive success. So here is another speculation. Language may have served as another "magic trait," which helped drive the later evolution of *H. sapiens.* A man, or woman, attracted to the sounds we know as language would have had children able speak better and more likely to mate with a good speaker. This could have driven the evolution of a human species specialized for language. Perhaps our ability to whisper sweet nothings into a lover's ear is what separated us from the Neanderthals.

I do not want to push this idea too far. Like much speculation about human evolution, it is difficult or impossible to prove. Yet, I believe, it serves a useful function, which is to emphasize that anatomical changes alone do not necessarily lead to new a species. We are not merely big-brained chimps. We find chimps funny and interesting, but we do not find them sexy. I suspect they feel the same way about us.

## Species of people

Years ago, when I first started thinking about human evolution, I searched the literature for field data on primate deaths. Hard facts were few. In the jungles of Africa and South America, where most nonhuman primates live, individual deaths are difficult to observe. Predation takes some individuals, but tree-living primates are well protected against most predators (other than modern humans). In one case, researchers were able to collect the carcasses of several spider monkeys that had died under natural conditions. Surprisingly, autopsies revealed that most had perished from toxins that are common in the leaves of tropical plants. For most other primates, and for our distant ancestors, nothing is known about the major causes of death, or reproductive failure.

Suppose, for arguments sake, that a major cause death of our primate ancestors was energy- sapping, disease-transmitting parasites. Anybody that has hunted, or has come across a recently killed wild mammal, can attest to the swarms of ticks and fleas they carry. These may not kill instantly, like lions and eagles, but they are killers nonetheless. Perhaps our wonderful hands did not originate for making tools, but for picking fleas, ticks and lice from our bodies. If nothing else, this might explain why we are nearly hairless. It makes finding these pests much easier. This hypothesis has about as much supporting evidence as those that propose that we began as big game hunters. I doubt, however, that we will ever see a popular book about human evolution called "Man – The Nit Picker."

Today, most biologists agree that people are all members of a single species, called *Homo sapiens*. All humans are potentially able to mate with any other human, which supports the one species view. Yet a biologist from another planet might argue with this conclusion, as sufficient morphological variation exists to justify dividing humans into multiple species. It would be fascinating to attend a conference of alien biologists at which they discuss the status of humans as a species. I imagine that the debate would be quite fierce. If I was participating, I would argue for a one-species model of the human population. This conclusion is based upon my definition of a species as: "A population of organisms with reproductive genes that function to preserve a set of harmonious ecological genes." The ecological genes that define humans include those that make our brains big and our hands good for making and using tools. While humans vary by region, we all depend upon learned behaviors (culture) for our survival, and can successfully interbreed.

Thus, humans comprise a single species. At least at the present time. However, this might be a temporary condition, and the species we know as *Homo sapiens* may be on its way toward splitting into multiple species. This prediction is based upon two considerations: Our unique ecological adaptations and our reproductive biology.

Consider the power of our brains and hands. These created an entirely new mode of ecological adaptation, and could possibly initiate a burst of "adaptive radiation." This term describes what takes place when a new adaptive strategy emerges, and leads to diverse new ways of using that strategy (it also describes the burst of speciation that can occur in an isolated environment, such as a lake or island). A good example is the diversification of birds. The first birds were presumably members of a single species. Feathers and wings were so useful, however, that variants on this theme quickly appeared. Soon the sky was filled with hundreds of species of birds. Similarly, our oversized brains and skilled hands may present a new opportunity for adaptive radiation.

In addition to their adaptive novelty, birds rapidly diversified for another reason. They tend to be highly selective in their choice of mates, a trait that promotes speciation. Humans, compared to other mammals, are also picky about their mates. Thus, we have two traits that should promote speciation – adaptive novelty and reproductive selectivity. When the time comes, as it ultimately must, when human population can no longer increase, when we have reached the Earth's carrying capacity for our species, evolution will work its magic. The changes that might occur, which will take many thousands of years, are unpredictable, but still fun to speculate about. (On a whim, I have penned a short story that illustrates these ideas, which I have attached as an appendix.)

I often ask my students to predict the future of human evolution. While they often come up with creative responses, they rarely come close to a scientifically valid answer. No, we will not evolve gills behind our ears (like Kevin Costner in *Water World*), nor are we likely to become super strong or smart. After letting students struggle with this for a while (go ahead, give it a stab), I use the exercise to reinforce the basic principles of evolution.

Evolution requires two things: Differential reproductive success, and genes that alter reproductive success. If these exist, then evolution by natural selection must necessarily occur. So, to make predictions about human evolution we need to answer two questions: Who is most likely to reproduce, and does their success have a genetic basis? A visit to the web site of "The Darwin Awards" (www.darwinawards.com) reveals many means by which people remove themselves from the gene pool, but these are not necessarily genetically based, or widespread.

As an evolving population, consider the United States in the twenty-first century. Here the most common causes of premature death (before the age of 21) are accidents and violence. An eighteen-year-old who dies in a speeding car, or a street fight, is unlikely to contribute genes to future generations. Ample evidence exists that high testosterone levels, a genetically based trait, promotes such dangerous behavior. High testosterone levels may have, at one time, enhanced reproductive success, but today seems likely to have a negative impact. So a few thousand generations of angry young men getting selected out of the gene pool will produce a nation of wimps – or maybe not. Nonaggressive men may be less likely to die violently, but they may also be less likely to be economically successful.

Speciation of sexual organisms also, of course, depends upon mate choice. Perhaps some women prefer to mate with nonaggressive men. Others like their men macho. The children of these matings will be fit, but in different ways that will depend upon their environment. What are the evolutionary consequences of this? At the present time, not much, other than to maintain the normal distribution of testosterone production. In the future, who knows?

## Why this matters

Some creationists argue that evolution cannot explain how one species emerges from another. While these creationists may accept the validity of microevolution, during which a population undergoes gradual genetic change, they refuse to accept macroevolution, the appearance of new species. They argue, incorrectly, that science has never seen one species turn into another. This is not surprising given how slow speciation is.

Heavens help us if the creationists discover that biologists cannot even agree on how new species form.

Oddly, I have never read a creationist tract that attacks evolution for its failure to explain sex. I guess that these highly religious folks are too embarrassed to bring up the subject. Few creationists will read this book, but others will find it of interest, and they will gain an understanding of how evolutionary biologists think. All too often, the debate about evolution gets bogged down in trivia. Creationists love nothing more than attacking evolution because of a perceived weakness in some set of biological data. The validity of evolution does not rest on the wings of the much maligned Pepper Moth, (which may, or may not, have changed in response to industrial pollution), but is firmly established by both theory and observation.

In spite of the power of evolutionary theory, major gaps in our understanding of life persist. Let's not waste our time arguing with creationists, who are clearly not interested in facts or logic, but we should certainly be more animated in our consideration of sex and speciation – those dilemmas that puzzled Darwin and continue to mystify us today.

# Appendex

## My Choice (a short story)

Dear mother, Gribitz and I will soon mate. This news may upset you, and I understand. Hybrid is a dirty word, and mating out of your own species is recognized as a sin. This is true for all human species, including Gribitz's. Yet, I believe that my choice could prove beneficial.

Let me explain.

Nearly a million years have passed since Darwin wrote The Origin of Species, but only in the past thousand years or so have we come to appreciate how evolution would change us. In Darwin's era humans were members of one species. This seems strange to us today. You might be surprised to learn that the archaic classic, Romeo and Juliet, is not about the tragedy of hybridization. Romeo and Juliet were of the same species, the now extinct *Homo sapiens,* but from different families.

I have spent many hours plugged into Google BrainMeld©, to trace the scientific study of the family *Homo.* At one time, just two human species, *H. sapiens* and *H. neanderthalus,* shared the planet. Scientists in the 21st century (C.E.) reported that *H. sapiens* and *H. neanderthalus* had, on occasion, engaged in sex. Oddly, gene flow appeared to occur only from Neanderthals into *H. sapiens.* Thus, *H. sapiens* carried *H. neanderthalus* genes (as we still do today), but the converse of this was not observed.

So we are even today a hybrid species.

Why the *H. sapiens* and *H. neanderthalus* hybrids thrived we will never know, but I have a theory – and it explains why I will mate with Gribitz.

The other day I shared a pot of stew with Gribitz. This may surprise you, as members of our species rarely share food with non-relatives. While we ate, we discussed our future, and possible fate of the children we might have together.

Gribitz is a marvelous dancer, but this is not why I choose him. He tells me that he still plans to attend occasional mating leks at the dance hall, and I'm ok with this. What matters most is that my own offspring will thrive. Motherhood is a great burden, and a woman wants the best father for her child. At one time men were also picky about their partners, and they also helped raise the children. This changed with the Great Collapse. Not only were most cities abandoned, but so too were the social structures that made them possible. Afterward, life became hard for all humans. Some retreated into the woods, and became hunters. Others stayed close to water and found sustenance there. Some maintained the historic traditions of farming, and scratched out a living from the depleted soil.

In this way, the Great Collapse began of a new round of human evolution, and the emergence of additional human species. It was an obscure biologist from Philadelphia who first predicted this might happen. He pointed out that the diversity of species correlates closely with mating habits. Millions of singing, dancing and brightly colored species should have made this fact obvious. Among most mammals, however, females have little choice in their partners. As a rule, when she is ready to conceive, she makes herself available to the dominant male, or the first male she comes across. This lack of choice makes it difficult for mammals to evolve the selective mating practices that help produce a new species.

Humans are different, and like many birds and tropical fish, we select our mates based upon genetic markers. Early in our evolution men did the choosing, and women are still burdened with the oversized breasts that once advertised our potential value as mothers. After the Great Collapse, women began to reassert their natural right to choose their child's father.

Another important change occurred. Humans, who were once generalists, became specialists. A warrior species, with enhanced strength and aggressiveness, was the first new version. Big aggressive women selected big aggressive mates, and produced big aggressive children. The brilliant red hair of *Homo macho* is unrelated to these adaptive traits, and it was by pure chance that it evolved as a visible marker of that species. Cardinals are red for the same reason.

Over many thousands of years other *Homo* species emerged, including our own, *Homo sarcophage*. Our exquisite sense of smell enables us to find food that other scavengers, and other human species, overlook. A deer rotting in the forest is a feast for us, if we can beat the vultures to it. At one time most people raised their food, a tradition continued today by *Homo agrigo*, although they barely survive the frequent raids of *H. macho*.

Agility also has its value, and the evolution of Gravlik's species, *Homo tango*, was the result. When Gravlik attends a mating lek he is often successful, and attracts an ovulating partner. Clearly he has "good genes" and I expect that he will pass some of these on to our children, but this is not why a choose him. A story from the Fossil Fuel Age tells of a famous scholar by the name of Shaw who was introduced to a beautiful actress. "Wouldn't it be great if we mated," she supposedly said to him. "Our children would have your brains and my beauty." His reply: "Unfortunately, madam, they would probably have your brains and my beauty."

So why would I toss the genetic dice and choose to mate with Gravlik? Our children may not dance like him, but I think they will be more agile than me. They may not have my keen sense of smell, but they should be better than some at finding precious protein-rich treats. Yet it is not these anatomical traits that matter. Something bigger is at stake.

The great cites whose ruins litter our landscape depended upon two things – plentiful energy and human cooperation. Energy went first, followed quickly by cooperation. The evolution of cooperation is a mystery, and theories dating from the Fossil Fuel era suggest that it is fragile, and can be easily destroyed by the appearance of cheaters. This, more than anything else, led to the breakdown of civilization.

I believe that we are beginning to see a return to the days of cooperation. In our own village people come together to build trails and latrines. Gravlik, even though he is not one of us, has helped. And I have gone to his village to teach their young to read, to help them appreciate the great history of the human family. We do this for each other.

I choose to mate with Gravlik because he will work with me to raise our children. In my dreams I see our children possibly contributing to the formation of a new species, one which may restore the grandeur of past civilizations. I have even given a name to this new species: *Homo amour* – loving man.

# About The Author

Richard C Weisenberg is a Professor Emeritus of Biology at Temple University in Philadelphia. His education was in physics (BA, University of California, Santa Barbra) and biophysics (PhD, University of Chicago). His ground-breaking research in cell biology has been published in major journals and presented at universities and professional conferences both in the US and internationally. While his research focus was in cell biology, he developed an interest in evolutionary biology, and taught this subject to biology majors at Temple University. It was from this experience that the ideas for Why We Have Sex emerged.

In addition to this book he is the author of the science-based novel The Doctors Cat, and the soon to be published novel Bird in the Window.

He lives in Philadelphia with his wife of 30 years. Feel free to contact him at rcw@temple.edu.

www.ingramcontent.com/pod-product-compliance
Lightning Source LLC
Chambersburg PA
CBHW051534170526
45165CB00002B/728